A healthy civilization can only be one that harmonizes with and integrates into the totality of life, enhancing not demolishing it.

JOSÉ LUTZENBERGER, BRAZIL (1926–2002)

First published in 2020 by Little Toller Books
Ford, Pineapple Lane, Dorset

Printed in the Wales by Gomer on Arctic Volume paper

ISBN 978-1-908213-83-9

 Some of the information captured in the accompanying stories may have changed since photographs were originally featured in the *We Feed the World* London exhibition, Autumn 2018.
All proceeds from this book go towards supporting food sovereignty.

Frontispiece:

Naturalezan Viva, Guadalupe Norte, Argentina (*see* page 124)
Photographer Jordi Ruiz Cirera

WE FEED THE WORLD

A celebration of smallholder farmers and fishing communities

Little Toller Books

CONTENTS

7 Introduction
Liz Hosken

11 Foreword
Vandana Shiva

15 About We Feed The World

17 Cooling the planet
49 Increasing seed diversity
75 Empowering women
93 Nutritious healthy food
119 In harmony with other species
143 Reviving community wellbeing
171 Custodians of the land and nature
205 Regenerating our soil
231 Protecting our water
253 Providing meaningful work for all

281 The Gaia Foundation
283 The Photographers
291 Acknowledgements

Opposite:
Naturaleza Viva, Guadalupe Norte, Argentina (*see* page 124)
Photographer Jordi Ruiz Cirera

INTRODUCTION

Liz Hosken

Welcome to this celebration of the small farmers and fishing communities who feed the world!

The stories you will find in these pages are first and foremost a testimony to the knowledge, resilience, dedication and skill of these farmers, fishers and their communities. Resilience is rooted in their intimate relationship with the land and waters across our planet, enabling them to read the laws and the cycles of their habitat and produce food in harmony with nature. Their dedication, sense of responsibility and love of the land and waters where they live have led them to protect, regenerate and sustain the wellbeing of the ecosystems on which we all depend. Their knowledge and skills, developed over generations, have enhanced the diversity of crops and wild species in our food system, and restored the health of the ecosystems that sustain their farming and fishing.

Their intimate understanding and respect for nature have inspired these and other farmers and fisherfolk to maintain a virtuous and age-old cycle of reciprocity. As one of the farmers in this collection, Ron Finley from Los Angeles, says: 'When you plant a garden, you're not just feeding people, you're healing the planet, the soil and yourself, all at the same time.' The El Choro community in Bolivia expresses it in this way: 'Sumak kawsay (their philosophy) taught generation after generation to live in harmony with themselves, with their community, and above all with Mother Nature.'

The poignant stories in this book also bear witness to both the need and the capacity of farming and fishing communities to resist the forces of industrial agriculture and fisheries. A young farmer, Zuzana Pastorková, from Slovakia, says: 'This is very different from industrial agriculture. To understand what's happening in nature you have to be quiet and observe and then the answers emerge on their own.' Colin Seis, from Australia, echoes this sentiment: 'Being a farmer now (having abandoned chemical agriculture) is very easy because I just

Opposite:
The Walronds, Glebe Farm, Somerset, England (see page 22)
Photographer Kate Peters

let Mother Nature run it for me.' Nature has an enormous capacity to heal if we work with her.

Resisting the dominant paradigm, while forging strong relationships with local communities and ethical suppliers, shows that another kind of agriculture is possible. As Rob Walrond, a farmer from the UK, says in his understated way, these local networks, united in solidarity, are 'a little more understanding than the supermarkets when a flood wipes out the onions'. Stories like this provide important lessons in the era of climate change, species extinction and global pandemics. They can guide us in how we can support resilient food systems, and what we should reject and why. We hope this book will inspire a commitment to support these heroic small producers.

We need to be alert to the latest narrative of the big companies who claim that they, too, are invested in regenerative and agroecological farming. Their PR is well spun and enticing. An unholy alliance of Big Tech and Big Ag is promising improvements in farming through 'precision agriculture' that will enable farmers to be more precise in the agro-poisons they use, as well as through the use of patented, gene-edited crops and other silver-bullet solutions. In reality, these measures mean further stripping farmers of their knowledge and autonomy, increasing their debt burdens and creating dependency on corporate technology, seeds and pesticides.

José Lutzenberger, a much-loved Gaia Foundation Associate, often referred to as the 'father of Brazil's environmental movement', was himself a former Bayer employee. He left the company in protest and became one of the world's leading critics of industrial agriculture. 'The modern farmer', said Lutzenberger, referring to those working on large-scale operations, 'is only a little cog in an enormous techno-bureaucratic infrastructure that requires special legislation and heavy subsidies. Compared to his predecessors, who did almost everything related to production, processing and distribution of food, he is not much more than a tractor driver and poison spreader.'

It is not enough to declare our opposition to this dystopian future for agriculture. We must make strong, tangible efforts towards seed and food sovereignty by actively seeking out and supporting smallholder farmers and fisherpeople. These are the people developing and maintaining regenerative food systems that heal soils, sequester carbon, value and maintain a diversity of crops and wild species, and protect community and farmer autonomy from corporate capture. We can support them by buying their produce, influencing friends and family and campaigning for a regenerative farming future.

We are grateful to the photographers who generously gave their time and their art to tell this story. And to Francesca Price and Cheryl Newman, the visionaries who brought together this celebration of food, farming and photography, alongside the Gaia team.

Most of all, we are grateful to the farmers and fishing communities whose dedication to their calling and love of Mother Earth helps feed the world and care for our living planet, in spite of the great odds they are up against.

We hope their stories inspire you to support the small farmers and fisher-people where you live, and to play your part in strengthening food sovereignty now and for future generations of all species.

Liz Hosken is co-founder and director of The Gaia Foundation. She is inspired by indigenous peoples in the Amazon to nurture 'affectionate alliances' to collaborate in reviving biocultural diversity and protecting sacred lands in the Amazon and Africa. A prominent advocate of Earth Jurisprudence, she works alongside partners and communities to restore the knowledge and practices which recognise that we humans are embedded in a lawful universe, and need to comply with these laws to sustain life on our planet. She supports the vital role of social movements in holding power to account and building a critical mass to shift from the dominant human-centred ideology to an Earth-centred relationship with 'Gaia', our Mother Earth, in all our human endeavours.

FOREWORD

Vandana Shiva

Food is not stuff. Food is not a commodity. Food is the currency of life. The food web is the web of life. Growing food according to the laws of the Earth regenerates the land and produces nourishment. Food as life and nourishment is real food, grown by real farmers – farmers like those whose stories are told in this book.

Industrial agriculture produces commodities, not food. That is why, as industrial agriculture spreads, hunger and disease spread and waste increases. According to the Food and Agriculture Organisation (FAO), a billion people are hungry, and another two billion suffer from food-related diseases.

Contrary to the fallacy that small farmers and their agroecological systems are unproductive and therefore dispensable, thus leaving our food future in the hands of the Poison Cartel, small farmers are providing 70 per cent of global food using just 30 per cent of the materials that go into agriculture. In direct contrast, industrial agriculture is using 70 per cent of the land to provide only 30 per cent of our food.

Commodity-based-fossil-fuel and chemical-intensive agriculture and food systems have contributed 50 per cent of the greenhouse gas emissions that are causing climate havoc, threatening agriculture. They have caused 75 per cent of soils to be destroyed, the destruction and pollution of 75 per cent of our lakes, rivers and oceans. They have pushed 93 per cent of crop diversity to extinction. Additionally, intensive industrial agriculture is also creating a health crisis through the production of nutritionally empty, toxic commodities and contributes to 75 per cent of food-related chronic diseases.

This industrial path of food and farming leads to a barren, dead planet, spreading poisons and chemical monocultures across continents. Children are dying for lack of food; debt-burdened farmers are committing suicide; people are dying from chronic diseases caused by toxic commodities devoid of nutrition and sold as food; and climate havoc is wiping out conditions for human life on Earth.

Organic farming takes excess carbon dioxide from the atmosphere, where it

Opposite:

Guillermo Ferrer
Sa Torre des Xebellins, Ibiza (*see* page 34)
Photographer Laura Hynd

doesn't belong, and through photosynthesis puts it back in the soil, where it does belong. It also increases the water-holding capacity of soil, contributing to resilience in times of droughts, floods and other climate extremes.

Chemical agriculture does not return the organic matter essential for fertility and for maintaining nature's life cycle to the soil. Instead it contributes to desertification and land degradation. It also demands more water, since it destroys the soil's natural water-holding capacity. Industrial food systems have destroyed the biodiversity of the planet both through the spread of monocultures and through the use of toxins and poisons which are killing bees, butterflies, insects and birds, leading to the sixth mass extinction.

As we look to the future, there are two divergent paths before humanity – one rapidly moving down an industrial path, promoting 'farming without farmers', and 'food without farms'. This path will uproot more farmers, destroy more forests and biodiversity and spread more food-related chronic diseases. The other is the agroecological path of food sovereignty. This path is being protected, nurtured and promoted by the millions of small farmers and fishing communities across the world who have kept their eco-friendly practices alive in spite of policies, subsidies, research and markets promoting inefficient, planet-destroying, health-destroying industrial agriculture.

Going further down the industrial path is a recipe for hunger, disease, climate catastrophe and extinction. Regenerating our planet through agroecology and food sovereignty will strengthen our rural economies and our health. The ecological path is imperative for the survival of our species.

Dr Vandana Shiva is one of the world's most dynamic and provocative thinkers on food systems, women and ecology. A scientist, activist and author, Vandana founded both Navdanya, a movement for biodiversity conservation and farmers' rights, and the Bija Vidyapeeth (School of the Seed) / Earth University, a centre of excellence for organic farming. Her books include *Water Wars*, *Monocultures of the Mind* and *Who Really Feeds the World?*

Opposite:
Selkie, Finland (*see* page 184)
Photographer Joel Karppanen

ABOUT WE FEED THE WORLD

This project started life as a photographic exhibition celebrating the triumphs and challenges of smallholder farmers and arguing for fair and just farming systems that regenerate the Earth. With the commitment of 46 world-class photographers, 50 communities and families, and hundreds of global partners and allies, The Gaia Foundation wanted to share the inspiring stories of those people who really feed the world. Through compelling images, the aim was to capture the social, political, ecological and economic contexts within which passionate and compassionate farmers produce over 70 per cent of the world's food. After three years, the project had compiled images and short stories illustrating how agroecological farming, as well as artisan fishing practices, not only provide the majority of the world's food but also offer solutions to many of our converging planetary crises.

We Feed the World opened to great success on London's Southbank in Autumn 2018; discussions around seed and food sovereignty and regenerative farming practices added to the reach and impact. Alongside this flagship exhibition, all those communities featured in the project had the opportunity to celebrate their achievements and engage local audiences by curating their own mini exhibitions with packs of printed images sent worldwide. These pop-up exhibitions took place in barns and libraries, shopping malls and village halls. They brought together farming and fishing communities geographically worlds apart but connected in their resilience and ecological wisdom.

The stories accompanying these striking images are evolving as the farmers and communities featured navigate new and emerging struggles. Whilst some of the stories may have changed since the flagship *We Feed the World* exhibition, the spirit of the work remains strong in each and every one. This compilation is a true celebration of the diversity of small-scale food and farming around the world in the twenty-first century.

With such an impressive body of work and critical message to share, it seemed a natural next step to adapt the exhibition for publication as a book that can be enjoyed for decades to come. In your hands, *We Feed the World* can continue to explore the future of food and farming, challenging the dogma that an industrialised food system is what we need to feed the world.

Opposite:
Cajamarca, Tolima, Colombia (*see* page 240)
Photographer Federico Pardo

The way our food is produced and ends up on our plates could hold the solution to the biggest crisis currently facing our planet.

COOLING THE PLANET

The industrial food system, which relies on chemical fertilisers and pesticides, heavy machinery, processing factories and long-haul transport, is responsible for at least a third of all our global greenhouse-gas emissions. This also takes into account the deforestation caused by clearing trees to grow vast monocrop plantations such as soy, sugar cane, palm oil, maize and rapeseed, which are used for animal feed and fuel and are found in most processed foods sold in supermarkets today.

In contrast, small-scale farmers around the world use regenerative agricultural methods, which not only reduce emissions but actually reverse climate change by drawing carbon away from the atmosphere and storing it in healthy soils, otherwise known as 'carbon sequestration'. One such method, known as no-till agriculture, involves planting seeds directly into the soil and leaving the organic matter from the harvested crops to fertilise the land and retain the moisture. This simple act enables the microbes in the soil to absorb carbon, potentially storing between 5 and 20 billion tons of carbon dioxide a year.

Supporting smallholder farmers who follow the laws of nature and apply agroecological approaches also ensures more trees, fewer chemical fertilisers and pesticides, shorter supply chains and less food waste, all of which helps to reduce carbon emissions into the atmosphere.

MZEE, LUKINDU, LWENGO DISTRICT, UGANDA

John 'Mzee' Ssentongo worked as a government accountant for more than 22 years in the Ugandan capital, Kampala, before returning to his ancestral land in Lukindu village. He was shocked to find it abandoned and the lush and beautiful landscape of his childhood now dry and barren. Drawing on traditional wisdom, Mzee started planting fruit trees: mangoes, papayas, avocados, oranges and jackfruits. Now, after a couple of decades, at 72 years old, Mzee is caretaker of a 'food forest' which hosts visitors from all over the world eager to learn from his farming philosophy.

Over the last 25 years, nearly 40 per cent of Uganda's forests have disappeared – prey to land-grabbing agribusiness plantations and housing developments. Mzee wanted to reverse this trend, and with the help of his two daughters and seven grandchildren, he has planted over 1,000 indigenous trees on his five-acre farm. He says, 'Each mango and jackfruit tree will live and produce for more than 100 years,' which is the kind of resilience and longevity we need in such climatically unstable times. Each tree will attract and maintain a bountiful water supply, making the surrounding land ready for growing other crops, even when rainfall is scarce.

Now under his cool canopy, Mzee cultivates a huge variety of vegetables and cereals, including indigenous greens, edible flowers, pea aubergines, watermelons, potatoes, rare banana varieties, coffee, herbs, flowers and maize. But the prize of his garden is, of course, the fruit, which he sells in Lukindu and the nearby Kiwangala market. Alongside the crops and other plants, insects and birds flourish in the garden and Mzee catches and eats the delicious crickets that take refuge here – a great source of nutrition.

Mzee sees planting trees as a way to build resilience to climate change and he teaches his methods to many farmers from the surrounding villages. Much of his fruit is shared with the schoolchildren who pass by each day, as he enjoys engaging the future generations and teaching them what the trees really mean. He says, 'Diversity is important because it allows plants to react to unexpected events, to adapt to climate change and to resist parasites and disease. A system based on a limited number of varieties, on the other hand, is very fragile.'

Photographer
Kibuuka Mukisa Oscar

KEEP UGANDA GREEN
PLANT MORE TREES
JACK-FRUIT

TULI BUMU
ASSOCIATION
TUDA
for the future

THE WALRONDS, GLEBE FARM, SOMERSET, ENGLAND

Glebe Farm on the Somerset Levels has been in the Walrond family for 200 years. Rob and Lizzie have seen first-hand the huge changes that have taken place in British farming, from the introduction of chemical fertilisers to the pressure of supermarket monopolies to what they say concerns them most – the extreme weather patterns that now dominate food production in the UK. 'There has always been unpredictability because it's the weather and we're in England,' says Lizzie. 'But in the past you might have had a ruined harvest due to flooding or drought once every 20 years; now it's happening far more often.'

Despite the challenges and its small size, Glebe Farm has continued to thrive and now offers an array of over 100 types of vegetables from its on-site farm shop. Lizzie and Rob put this success down to three things: the organic system they converted to 20 years ago, the huge diversity of produce and the decision to sell direct to their customers, 'who are a little more understanding than the supermarkets when flood wipes out the onions'. As Lizzie says, 'It can be more work, managing many different activities such as planting trees, hedgerows and medicinal herbs for the sheep, but it has allowed us to work with nature in a woven interplay of predator and pest that is more resilient to extremes.'

These changes in the weather are affecting the whole community. Long, wet winters are delaying planting, which means the 'hungry gap' (the traditional period in spring when there is little fresh produce) is getting longer. Meanwhile, hotter summers leave crops like barley struggling for moisture, and brassicas are often wiped out by pests that would normally be killed off by the first frosts. 'It's not a healthy pattern,' says Rob, 'and you never know what you're going to get next'. In 2017, the Walronds' neighbours brought in the last wheat in early December – unheard of in the UK where the traditional wheat harvest is August.

Rob believes that we all have a role to play in the stewardship of our food and farming system, especially with climatic instability. 'It's the most important thing for our health and our life, yet we degrade it and don't value it properly. Farmers do their bit by producing the food but we can only produce what people will buy. At the end of the day, whoever buys the food has the final say.'

Photographer
Kate Peters

THE MAZVIHWA COMMUNITY, ZVISHAVANE, ZIMBABWE

The farmers of central Zimbabwe know all too well the impact of climate change. Zimbabwe was once known as the breadbasket of Africa, but the erratic rainfall has left the country struggling to grow food and is pushing the harvest back later every year. Many crops die, waiting for water.

The crisis has been made worse by the fact that most farmers have been growing government-promoted hybrid crops, which are especially vulnerable to climate fluctuations and require chemical fertilisers that dry the soil.

These factors have led many communities to turn their backs on these hybrid maize varieties and return to the indigenous seeds that their ancestors once grew; seed varieties that have adapted over time to withstand long periods without water and don't require chemical inputs to thrive. One of these communities is Mazvihwa in south-central Zimbabwe; they are now reviving traditional crops such as rapoko (finger millet), bulrush millet and sorghum.

For these farmers water is the single biggest issue as the rainy season gets later every year. The schoolchildren in Mazvihwa remember finishing their exams in October, ready to go home and plant, but these days it can be February before the rains have even arrived.

To help conserve as much water as possible, the community has adopted the water-harvesting techniques of the late farming guru, Zephaniah Phiri. A local farmer, Phiri was revered across this region of Africa for his techniques which involved building terraces and utilising the natural contours of the land to keep water in the fields and nourish the soil. Mazvihwa farmers now build 'phiri pits' to collect and store rainwater and some even have ponds where they farm fish amid this arid landscape.

The changes have produced significant results for Mazvihwa. Farmers, who once produced one tonne of grain, now harvest four tonnes. The chief of Mazvihwa has even been nicknamed the 'chief of small grains' because of the plentiful local harvest. These days, the farmers photographed here, Austin Mugiya, Forward Zivanemwoyo, Benedict Muzenda and Maria Fundu, uphold the tradition of *nhimbe*, each taking it in turn to help each other with their harvests.

Austin says, 'when it comes to farming, we found out that it is not only in our minds but in our blood. We feel happy to make our own healthy food and be masters of conserving our environment.'

Photographer
Pieter Hugo

GUILLERMO FERRER, SA TORRE D'ES XEBELLINS, IBIZA

Guillermo Ferrer's 17-hectare farm, Sa Torre D'es Xebellins, is situated in the south-east of Ibiza, just outside Ibiza town. Guillermo was born on the farm in 1956, at a time when they had no electricity, bathroom or cars. 'We had candles, hot water from the wood fire, two horses, two mules and a chariot,' Guillermo fondly recounts. 'There were pigs, goats, sheep, chickens, turkeys, ducks and bees and a pond teeming with frogs, fish and all kinds of life – a real paradise for a child. At any time of the day, there was a magic calm between humans and nature.'

After some years of travelling, Guillermo returned to the farm. But what he met with was far from his childhood memories of abundance. Instead, the land had degraded into a desert-like panorama, with dying trees and almost devoid of life. Tourism had taken off in Ibiza and farm after farm had been abandoned in favour of hotels and infrastructure. Guillermo's family farm had been completely abandoned. There were no animals, no gardens, no flowers and no fruit trees. All that remained were the hardy, 1,000-year-old olive trees, providing the last shade on the farm.

As Guillermo sat beneath the olive trees and the hours passed, a vision came to him that he must follow the tradition of his ancestors and create once more not just a farm, but an oasis of life. The early years were the hardest because the insects and wildlife had all but left, and so the land felt lifeless. But he distinctly recalls the day, four years after starting the regeneration, when a ladybird reappeared. This was a moment of celebration and signalled the change in fortune for the land.

Now, 35 years later, his farm is buzzing with the sound of bees, with over 40 healthy beehives. In turn, the insects have attracted the birds, and with the birds have come the birds of prey. He is proud that the island's few eagles are regularly seen circling the skies above his rich land. Guillermo's land boasts over 300 fruit trees and feeds over 250 local families who visit the farm directly to collect their fruit and vegetables, eggs and honey.

'The deep vision is only to create life. The big old original trees create the habitat. The fruit trees and vegetable gardens give the food. The colourful flowers give joy and attract the bees. Around 300 chickens sleep in the trees, feeding the earth with their shit. Ducks, turkeys, birds, bees, insects live in harmony with the wild animals. A healthy ecosystem expresses life and beauty through all the seasons, every day an abundance.'

Photographer
Laura Hynd

PUERTO COLOMBIA, AMAZON, COLOMBIA

The community of Puerto Colombia lies on the Tiquié River in Vaupés, one of the most remote regions of the Colombian Amazon. There are no roads here and the only way in or out is by boat. The indigenous peoples who call Vaupés home live and thrive through a combination of hunting, gathering, fishing and small-scale farming.

Six families live in Puerto Colombia, and each has several 'chagras', or forest gardens, where the community grows over 70 varieties of cassava, a dozen types of plantain, six varieties of pineapple and other tropical fruits endemic to the Amazon region. They also collect edible insects and mushrooms, which complement the wild game and fish to be found in the forest and its rivers.

Families create their chagras by felling trees to open up a small clearing in the forest that allows sunlight to reach the forest floor. Crops are planted here for three to five years before the garden is moved to another area, allowing the rainforest to regenerate naturally. By rotating the location of their chagras, the forest farmers of Puerto Colombia mimic the natural cycles of decay and regeneration that occur when big trees fall in the forest. In this way indigenous people in the Amazon have contributed to the biodiversity of the forest over thousands of years.

This way of farming is undertaken with a deep respect for the forest as a source of life and place of origin. For the indigenous people the forest, and all the beings that inhabit it, are expressions of an original vital energy that imbues all of life. Through the wise council of the shamans they seek to maintain a dynamic balance between the world of the forest, the world of the ancestors (the spiritual world) and the world of humans.

Passed down the generations, the shamans' deep knowledge of the forest has traditionally enabled them to guide their communities to make wise decisions about where and when to plant, hunt, fish and hold ceremonies. But today, with the changing climate, the shamans say that they are struggling to interpret the signs that have always guided them. 'It feels like the sun is getting closer to the Earth,' says one elder from Puerto Colombia.

As the rains arrive late and become more erratic, the farmers are working to adapt to this 'new normal' by adjusting their food-growing practices. Meanwhile the shamans are focusing their abilities ever more rigorously on communicating with the spirits of the forest in order to navigate the challenges of an increasingly unstable planet.

Photographer
Stefan Ruiz

KEEP
CALM
AND
BUYNEW
SHOE

Small-scale farmers increase seed and plant diversity in order to enhance productivity and resilience to climate change.

INCREASING SEED DIVERSITY

It is estimated that the diversity of three-quarters of the world's crops declined during the twentieth century. Seeds that have been locally adapted to withstand drought or floods, those that perform better in hot or arid climates or that are naturally resistant to pests disappear every day. Today, nearly 86 per cent of the food produced by commercial breeders comes from just 16 plant species. With such a narrow range of options to draw from, the global food system is extremely vulnerable, especially in the context of climatic instability.

Big agribusiness corporations promote their commercial seeds as the solution to feeding the world by promising that they offer higher yields. However, these seeds do not reproduce and farmers are forced to buy new seeds each year, alongside the chemical fertilisers that the seeds are bred to rely on. Millions of farmers around the world have become indebted to these companies, losing their farms and sometimes taking their own lives because they have nothing left. Also, over the years, the yields of these seeds decline as the fertility of the soil is compromised as a result of excessive pesticide use and other extractive farming methods.

In contrast, small-scale farms still hold three-quarters of the seed diversity of our global food system, which they save, share and trade locally. Unlike large monocrop plantations, which farm one singular crop variety on huge tracts of land, small-scale farmers rely on diversity, planting numerous different varieties and crops, thereby spreading the risk of crop failure. These seeds have been bred and passed down through generations, enabling them to adapt to their local environment and its particular challenges.

For indigenous and farming communities, seed is sacred, the destiny of each life encoded in its tiny frame. Seed is at the heart of their traditions, from rites of passage to rituals at sacred natural sites. Nurturing diversity is embedded in their farming practices – in the knowledge that in diversity lies the core of resilience.

DR DEBAL DEB, ODISHA, INDIA

Dr Debal Deb is a zoologist-turned-farmer and rice conservationist. On his small farm in India's eastern state of Odisha, Debal is preserving some of the most saline- and flood-tolerant varieties of rice in the world. A lone scientist with no financial support, he is determined to protect India's genetic wealth against corporate interests. The diversity of crops that Debal is now conserving is a critical contribution to the future of food and farming in an increasingly unstable climate.

It is estimated that India has lost as many as 110,000 local varieties of rice since the Green Revolution pushed commercial hybrid varieties and chemical-intensive agriculture on India's farmers. Today, only around 6,000 varieties remain and fewer are being grown every year. With each generation the knowledge of how to grow these traditional, locally adapted varieties is being lost.

To date, Dr Debal Deb has cultivated 1,420 rice varieties on just two acres of forested land in the Niyamgiri hills. Some of the varieties in Debal's collection have the ability to grow for months under nearly 4 metres of water, while others can tolerate high salinity. With the help of the local farmers, Debal operates his own seed bank, Vrihi – the Sanskrit word for rice. Unlike modern seed banks that preserve varieties in refrigerated vaults, inaccessible to ordinary farmers, Vrihi is a living community seed bank where local farmers bring rare seeds, which Debal then grows and redistributes in one-kilogram packets.

'The farmers take the seeds on the condition that they bring some back', explains Debal. 'They must return at least half a kilogram as proof they have cultivated it. Most give 1 kg to other farmers so the cycle continues. In seven years in Odisha, 4,200 farmers from six States have received the seeds and 350 varieties have been distributed.'

Vrihi ensures that the diversity of varieties that have been developed over centuries by local farmers in order to survive in marginal environmental conditions can continue and thrive. By contrast, modern agriculture relies on ever fewer varieties, making it far more susceptible to large crop losses. Indeed it was the locally adapted salt-tolerant rice varieties that were the only ones to survive when cyclone Aila struck the region in May 2009.

While Debal works tirelessly, he points out that 'after 60 years and billions spent on gene mining, the GM industry still doesn't have a single variety that can withstand a drought or seasonal flood or seawater incursion. But all of these characteristics are available in many of our farmers' varieties.'

Photographer
Jason Taylor

SHASHE COMMUNITY, MASVINGO, ZIMBABWE

In the dryland area of central Masvingo province, Zimbabwe, lies the vibrant community of Shashe. Until land reform in Zimbabwe 20 years ago, these 15,000 acres of land belonged to just three commercial ranchers who raised cattle for export. Today, it is home to nearly 400 families who farm a wide variety of grain, vegetables and livestock that give them enough to live on in a climatic and politically unstable environment.

Collectively, the community have recovered many ancient, indigenous varieties of crops such as millet, sorghum and maize, which are specially adapted to this dryland climate, as well as *tsvoboda*, sesame, amaranthus and many other local vegetables and fruit. They also keep goats, sheep, donkeys, pigs and chickens. Of the four ceremonies that mark the agricultural year, the biggest is the Matatenda festival, when all the families from the region gather for a week to celebrate the harvest and give thanks to the seasons and the spirits of the rains. A key part of the ceremony is the sacred millet beer, which is prepared by an elected group of women over a fire for a full seven days. Photographed here are Vongai Dube, Tendai Nago, Mhovai Matombo, Eucitina Wandai and Letticia Chisweto.

When the beer is ready the community gathers to partake in a ritual, which allows them to communicate with the spirits and mediate their relationship with their ecosystem. A procession is led into the forest where they offer a calabash of beer to the sacred Mubvumira tree. Music, singing and dancing follow, and the rest of the beer and harvest food are enjoyed. This is also a time when people come to share their seeds, farming testimonies and knowledge and experiences, helping each other improve their farming methods. Shashe is also the birthplace of ZIMSOFF (Zimbabwe Smallholder Organic Farmers Forum), which now represents 19,000 farmers across Zimbabwe. The Shashe Agroecology School trains farmers from far and wide, teaching inter-cropping techniques, water harvesting and regularly organising workshops, farmer-to-farmer exchanges, seed swaps and food festivals.

Nelson Mudzingwa, the national coordinator of ZIMSOFF and a resident of Shashe, says, ‘Achieving seed sovereignty is a passport to reach food sovereignty and our way of nurturing Mother Earth. Reconnecting to past experiences and practices such as rituals and ceremonies guarantees better communication and relations between the living and the spiritual world and helps us to navigate the changing climate.’

Photographer Jo Ractliffe

UNZWE
UTAPIRA KOITA
SHAGI

LIKOTUDEN, EAST FLORES, INDONESIA

Community leader Maria Loretha spent months travelling around the remote villages of East Flores talking to elders, before she eventually found the indigenous sorghum seed varieties that used to grow prolifically in this region of Indonesia. The ancient crop – now known for its superfood qualities – had all but died out on the volcanic island as successive governments encouraged farmers to grow commercial white-rice varieties and maize instead. Sorghum became known as a 'poor' crop, only suitable for feeding to animals.

However, these commercial varieties did not work in East Flores; the volcanic rock and lack of rain owing to changing weather patterns were unable to support the same wet-based agriculture that flourished in other parts of Indonesia. Despite increasing amounts of chemical fertilisers, for which the community had to pay, successive crops failed, and local families were left hungry, in debt, and faced with the prospect of having to leave and become migrant workers in order to survive.

In response to this dire situation, Maria Loretha began to mobilize the women of the Likotuden area to plant 30 acres of sorghum using the old seed varieties she had collected from the elders. The crop is more labour-intensive than rice and maize, but it requires less water and is more nutritious and versatile than these other grains. 'We all know that when we eat sorghum, we feel fuller for longer than when eating white rice,' says Maria. 'And it can be cooked as a porridge, made into a flour, cooked into brownies, pizza or a pop-sorghum, like popcorn!'

The experiment, which initially involved 62 families, has proven so successful that it has now expanded to other parts of Indonesia. For the women of Likotuden, sorghum has become a route to independence, allowing them to break free of a reliance on chemical fertilisers and pesticides, the devastating impact these have on the soil, and a cycle of debt. 'My friends say I am the maestro of sorghum, a master sorghum grower, but I am just an ordinary farmer,' says Maria. 'However, I do feel I live an extraordinary life while I live in Flores'.

Photographer
Martin Westlake

ISLES OF UIST, OUTER HEBRIDES, SCOTLAND

The Isles of Uist lie off the coast of Scotland, on the westernmost fringe of Europe, forming the last stronghold of both the Gaelic language and a crofting tradition that has maintained small-scale farming for generations. The farmers here pride themselves on being self-sufficient and not dependent on the Scottish mainland. And the brutal Highland and Island Clearances, 'the eviction of the Gaels', that saw tenants driven from their land in the eighteenth and nineteenth centuries are still remembered today, even among the youngest generations.

On Benbecula, Neil and Morag McPherson are third-generation crofters who grow small oats and bere barley for seed. Like most crofters on Uist, the family save seed from one year to the next, ensuring the local community can rely on the resilience of their crops which are adapted to the harsh climate of the Western Isles. There are 34 crofts in the village of Liniclate and at harvest time the community comes together to swap tools and machinery in return for labour, a tradition that goes back generations. For the families here, the crofting way of life is a source of great pride, as are the skills involved in their seed saving heritage.

Today, the crofters are working with The Gaia Foundation's Seed Sovereignty Programme across the Highlands to support the revival of heritage grains such as 'Eòrna', a traditional Scottish barley. The aim is to reintroduce heritage grain and grain growing in five Highland communities, from Lochaber to Assynt, and to revive the lost skills associated with grain growing.

Meanwhile over on North Uist, Angus MacDonald operates a herd of 300 highland cattle, which are fed on a mix of arable crops grown on the croft for winter feed. Years of hard work have made him the largest breeder of organic highland cattle in the UK. In autumn, Angus takes the animals over to the neighbouring tidal island of Vallay, where they will stay and graze for the winter. Angus's croft was handed on to him some years back by his mother Ena, who is now 78 years old and lives next door. She is well known nationally for speaking out for crofters' rights.

Ena still remembers the days of growing up here as the youngest in a busy family of five: hunting, fishing and helping the community harvest every September until nightfall. They had three cows that ensured there was always fresh milk and butter. Every Christmas they would kill some hens to send to their cousins in Glasgow, where food wasn't quite so abundant. Ena spent five years in Australia before returning to the croft to work alongside her father. 'You have hard times and you have good times. If you really love it you just carry on. It is something that is in your blood.'

Photographer
Sophie Gerrard

EL VISO DE LOS ROMEROS, MÁLAGA, SPAIN

In the Sierra de las Nieves mountain range of Andalucía, Alonso Navarro cultivates fruit, vegetables, cereals and medicinal herbs on 100-year-old terraces that make up his small farm, El Viso de Los Romeros. Despite the Spanish heat, there is no shortage of water which gushes along the ancient internal channels from the Jorox mountains. Against this lush backdrop, Alonso runs a small seed business and is president of the Andalusian Seed Network, a large network of 160 dedicated seed-savers who are bringing back traditional and lost varieties of vegetables and fruit.

Unlike most people today, Alonso has always been connected to seeds. He grew up in rural Andalucía in a traditional peasant farming family and learned to save seed from his grandparents after school. Each autumn they would roll the seeds up in tobacco papers, put them into envelopes and store them carefully, ready for sowing the next year. In those days, there was no such thing as seed companies. Family and friends would meet once a year to swap their seeds. 'I'd get radish seeds from my uncle, melon seeds from my cousin and *camorra* beans from the other side of the family.' Before seed became commodified, selecting, saving and sharing seed was an inherent part of farmers' knowledge and practice.

Today, corporations rule the seed industry in Andalucía, as they do everywhere else in the world. The successes claimed by advocates of the 'Green Revolution' have conversely diminished seed diversity, wiping out 90 per cent of Andalusian vegetable varieties and much more. When Alonso realised this 14 years ago, he decided to devote himself to preserving ancient diversity, the key to adapting to climate change. He sees seeds as the legacy of our ancestors. 'From the innocent practice of harvesting, we conserve a whole culture in our own hands. This knowledge was passed down to us from our grandparents and these seeds are transitory, passing from generation to generation, in a journey through history.'

Together with the seed-savers' network, Alonso holds an impressive seed collection: 700 tomato varieties (200 indigenous to Andalucía), 100 peppers, 80 beans, 60 cereals and more. Alonso and his grassroots team of growers are constantly in search of old varieties, testing them for favourable characteristics. Alonso found the pictured wheat variety, Recio de Ronda, in the village of Vega de Antequera. Now, revived by Alonso, Recio de Ronda is grown throughout Andalucía and is used in many bakeries.

Photographer
Spencer Murphy

TRIGO DURO
Guten Appetit!
ALFALFA

excelsior
just
for you
Ciezana

Women produce more than half of all the food that is grown globally. In the Global South, they are responsible for about three-quarters of the food produced.

EMPOWERING WOMEN

In most traditions around the world, women are the custodians of seed and farming. They are responsible for the food produced for the family, markets and community ceremonies. The diverse range of cultural foods that we enjoy today have been cultivated by women over generations, selecting and breeding seeds to have different traits, thereby enhancing the range of options they can draw on. They are highly ecologically literate, able to assess what to plant by reading the complexity of soils, the climate and the seasonal changes. In the context of growing climatic instability, their intimate knowledge and skills are essential.

Yet the industrial food system has systematically undermined the role of women, promoting cash crops and chemical inputs to men, who then use more land to pay back the debts they incur. This vicious cycle has squeezed women off the land and undermined their central role in traditional agriculture.

Today, thanks to farmers' movements around the world, the critical role of women in farming is being recognised and they are regaining their status and their land.

HOUENOUSSOU COOPERATIVE, TODEDJI, BENIN

Every Wednesday morning in Todedji, members of the women's cooperative (Houenoussou) gather at the house of their president, Salako, to eat together before heading off to their two-hectare market garden on the banks of the river Noire. Their work not only provides food for the village but it ensures that the traditional knowledge and ancestral seed varieties – which are more resilient to climate change – will be preserved and handed down to their daughters.

Just behind their garden lies the sacred forest of Oro. This is where some of the traditional ceremonies of the community take place, including those to call the rains and give thanks for the harvest. Every three years, in August, the community gather to sing and dance and pay their respect to the forest divinity, Oro, as required by tradition. After the ceremony is finished, the whole community is invited to join the festivities at the women's market garden, to feast on the fruits of their labour and to give thanks to all the other entities in the web of life which support them.

Forests such as this one are increasingly under threat in Benin from logging and commercial interests. Community and civil society leaders worked together to get them protected and in 2012, Benin became the first country to pass a Sacred Forest Law, recognising indigenous communities as custodians of these special places. Oro is now legally protected as a no-go area for hunting, logging, mining and any commercial or development activity. It recognises the custodian community's rich relationship with sacred forests and their ancestral lands which they have preserved for millennia.

Houenoussou, which means 'It's time to take destiny in our hands', provides a continual stream of healthy organic vegetables for the community all year round and their produce is increasingly sought after in the big cities where good quality produce seldom exists. Their children learn to grow food with their mothers from a young age, and help with the precious task of nurturing their crops. The women exchange seed and maintain a living diversity of varieties of a range of crops such as okra, cowpeas, and gboman (like a tomatillo). They also produce gari (manioc flour) and akassa (cornmeal) for the markets, which brings with it economic independence.

The communal market garden is a special place for the women, where they can be at ease, masters of their work and through their union, feel stronger. Collectively the women manage their 'Credit Union for Change' where they take turns to support each other to invest in ways that will support them, such as tools or more animals. Salako says 'we like to work in a cooperative because working together is fun. It is nothing more than pure synergy and solidarity. It allows us to be self-reliant in so many ways. We help each other and lead our lives together in understanding'.

Photographer
Fabrice Monteiro

EMBU COUNTY, MOUNT KENYA

The seven women shown in these photographs are small-scale farmers from the Embu County on the slopes of Mount Kenya. Cultivating between a quarter of an acre to three acres each, they produce a huge range of vegetables and fruits such as avocados, passion fruit and tree tomatoes. They also raise animals, grow tea and produce their own firewood and compost by integrating more trees into their traditional farming system.

Since the 1970s, Kenya has suffered massive deforestation as a result of mismanagement, illegal logging, charcoal production and the reallocation of forest land for the production of cash crops. In the last 15 years alone, Kenya has cleared land equivalent to the area of Greater London, Berkshire and Surrey combined. Together with climate change, this has led to reduced rainfall and increasing soil erosion which, in turn, has caused food scarcities and economic hardship for the communities here. If left unchecked, Mount Kenya Forest, which the mountain presides over, could lose much of its biodiversity and water-catchment value in 20 to 30 years' time.

The older women such as Agata, Lucy, Dionisio and Mercy still remember when their family farms were filled with trees and you could not see your neighbour's house because of them. Back then, cutting down a tree was considered a taboo and they had seasonal rivers that flowed through the land as a result. 'The rain was sufficient and the harvest was abundant,' says Dionisio. 'There was always enough to fill a grain store with food to act as a reserve during a drought period'. Today, the women are all too aware of the much smaller yields and the water crisis facing Kenya. They are working with International Tree Foundation on the '20 Million Trees for Kenya's Forests' campaign to address these challenges.

The women use agroforestry techniques to farm, planting more trees on their land as well as other crops. These trees can be pruned and used to feed animals or provide fruit or firewood which takes the pressure off the indigenous forests – and reduces women's workloads. Their canopies provide cover for other crops and the soil while helping to improve its fertility through the organic matter that drops to the ground. Meanwhile, the roots of the trees ensure water reaches deep into the soil and helps prevent erosion. It is a system that integrates more trees into traditional agricultural farming techniques and is already beginning to return these farms to the lush, food-filled landscapes they once were.

Entitling his photographs *The Seven Graces*, the artist refers to a trio of goddesses in Greek mythology who were associated with fertility and nature and have been represented in art throughout the centuries, often as allegories of nature.

Photographer
Omar Victor Diop

Tuko Pamoja

THE GODDESSES, ESTELÍ, NICARAGUA

Reyna Merlo and Isabel Zamora first met at a women's gathering in a makeshift tent 20 years ago. Isabel had endured many years of violent abuse from her partner and Reyna had been abandoned by her husband. As single mothers without land or money they had no choice but to work for the tobacco plantations that dominate Northern Nicaragua, and suffer the respiratory and skin diseases that result from the daily use of chemical pesticides. Their chance meeting led to a brave decision – to take control of their lives by creating an organic coffee-growing cooperative called La Fundación Entre Mujeres (FEM).

The cooperative, which also calls itself Las Diosas (The Goddesses), now has 1500 female members and provides a huge infrastructure for women in similar situations; they run their own schools, self-defence groups, seed reservoirs and a coffee factory. In recent years, they have diversified from coffee to growing vegetables, hibiscus, wine and honey, which has helped them to build the cooperative's resilience in uncertain financial and climatic times.

Through farming and food production, the women at FEM now have the economic freedom to make their own decisions about their education, their work and their bodies. They are no longer forced to rely on men for their financial security. FEM pays the women directly, enabling them to feel their efforts are reciprocated. 'The first time a *compañera* in our cooperative took home nearly 50,000 cordobas (US$1,923) as a result of her work, she couldn't believe it,' says Isabel. 'It was a sum that she'd never had in her life.'

Isabel did not learn to read or write when she was young, but now she holds a diploma and travels globally teaching agroecology. Reyna is one of FEM's seed guardians, who nurtures ancient maize and other varieties that are naturally resistant to pests and drought. She considers 'inherited food as a living being' and spends her time passing on her knowledge to the next generation of women, many of whom are now going to university. 'I sell my own produce now, I write on political issues and I teach. People know me. I have learned to be a visible, not an invisible, person,' she says proudly.

Photographer
Susan Meiselas

Good food is the cornerstone of good health. As the world faces global health epidemics, ranging from obesity to antibiotic resistance, the way we produce our food is under scrutiny.

NUTRITIOUS HEALTHY FOOD

The overuse of chemical fertilisers and pesticides in industrial farming systems has been linked to a range of human health problems, such as Alzheimer's disease, birth defects, cancers and development disorders. In addition, deforestation contributes to the emergence of novel pathogens like Covid-19. The regular use of antibiotics in factory farms, where animals are kept inside in cramped and often dirty conditions, has also exacerbated the problem of bacterial resistance to antibiotics. This is now being recognised as one of the biggest threats to global health and food security.

According to the World Health Organisation, unhealthy nutrition patterns, combined with physical inactivity, are responsible for the global obesity epidemic. In America, the current generation of children may be the first in two centuries to have a shorter life expectancy than their parents, owing to what they eat and drink. While in the UK, poor diets are estimated to cost the NHS £6 billion pounds a year.

In contrast, food produced by ecologically friendly farms have been shown to be healthier and more nutritious. According to research from Newcastle University, organic milk and meat contain around 50 per cent more omega-3 fatty acids than non-organic, as well as containing slightly higher concentrations of iron, Vitamin E and some carotenoid. Organically produced crops (cereals, fruit and vegetables) contain up to 68 per cent more antioxidants than non-organic.

How we farm affects the quality of the food we eat. How we shop does, too. Supporting small-scale farmers to produce a diversity of nutritious, affordable food will ensure that our good health and that of our families will continue for generations to come. In the words of Hippocrates: 'Let food be thy medicine and medicine thy food.'

GERALD MILES, CAERHYS ORGANIC FARM, ST DAVIDS, WALES

At the age of 16, Gerald Miles came home for the school holidays to help with the potato harvest on the coastal farm that his family leased on the westernmost tip of Wales. Months later, he was in the local bank manager's office taking on the mortgage – school was finished, he was now a farmer! The same determined spirit led him, many years later, to spend a week driving his tractor to London to protest against genetically modified foods.

Over the years, Gerald has become a master of innovation – critical in an area where family farms have disappeared as supermarket control impacts markets. When the potato market crashed in 1985, he was left with nothing and changed the farm to produce a greater diversity of crops. Ten years later, he converted to organic farming methods and sold milk, beef, corn and other cereals, and later re-invented it once more – this time as Community Supported Agriculture (CSA). He was always determined that the land continue to produce food and not be turned into yet another 'ghost village with holiday cottages'.

Today, Caerhys Farm sells a wide range of vegetables direct to sixty families in the area and has become a sought-after destination for young volunteers ('wwoofers') around the world looking to learn about organic agriculture. 'Having a CSA on the farm gives you a closer connection to the community,' says Gerald. 'Our aim is to produce food for the community in an organic way that takes care of the land. And that's what a farmer should be doing – feeding the local community with diverse, nutritious food.'

Spurred on to bring more cereal diversity to the farm, Gerald is now working closely with The Gaia Foundation's Seed Sovereignty Programme to run the Llafur Ni (Our Cereals) network. The group is reviving rare Welsh landrace oats and grains to share. They focus on intergenerational knowledge exchange, with older farmers accompanying new entrants to Welsh cereal farming, sharing their knowledge of traditional ecological growing practices.

Gerald has started growing ancient Emmer and Einkorn on his coastal fields. He knew that his father and grandfather grew a Welsh black oat on his farm, but he had lost the seed to continue cultivating this traditional crop. After years of searching, Gerald found a farmer in Ireland growing black oats and was able to have some sent to him in the back of a rugby tour bus. The black oat has returned to the land once again.

Photographer
Clare Richardson

VIVERO ALAMAR COOPERATIVE, HAVANA, CUBA

Between the Spanish-era colonial buildings and the post-revolution communist blocks of Eastern Havana, lies one of the biggest organic urban farms in Cuba: the eleven-hectare Vivero Alamar Cooperative. Here, José Manuel (former sailor), Fradel Martinez (former tobacco worker), Juan Portal (former petrol worker) and Juan Ramón (former fisherman) are among the 150 workers, all on equal wages, who meet each day to eat a freshly harvested lunch together before beginning their work as farmers.

Twenty-five years ago, Cuba was forced to face a challenge that most countries fear: life without oil. As a communist state, led by the revolutionary Fidel Castro, their main trading partner was the Soviet Union. When it collapsed in 1991, all Cuba's imports disappeared overnight, including food, petrol and agrochemical fertilisers. In this, the 'special period' that followed, Cubans had to survive on half the amount of food they previously ate. They also had to learn how to grow food without commercial fertilisers or machinery, both of which required petrochemicals.

Urban farms, called *organoponicos*, sprouted up across the city to fill the food gap, offering to retrain many who were now out of work. Vivero Alamar was set up in 1997, initially as an 800-metre-square farm, which used agroecological methods such as crop rotation and composting to produce food. Today, they harvest 300 tons of vegetables annually, most of which is eaten within the Alamar district – an area that previously had no fresh produce.

Without machinery, oxen became an essential part of the farm and remain so to this day. The workers – many of whom are senior citizens – produce a range of fresh vegetables and fruit including lettuce, herbs, beans, tomatoes, aubergines, mangoes, guavas, bananas and even mushrooms. Today, almost 90 per cent of Havana, a city of two million people, is fed on organic food produced by 4,000 or so *organoponicos* situated within the city limits.

José Monjes, one of the cooperative's agronomists says we must start to think about land and food in a different way. 'Land is not inherited from our parents, but borrowed from our children. If everyone only thinks about themselves then we will reach the point when there will be no planet to look after. Agroecology is nothing new, but it is a science and represents progress into producing food with minimal impact on the environment.'

Photographer
Michel Pou

SOUTHERN ROOTS ORGANICS, DORSET AND DEVON, ENGLAND

Dee Butterly and Adam Payne are part of a movement of young new entrant farmers who are returning to the land with the intention of making a social and environmental difference. At just 27, they made the decision to set up Southern Roots Organics Community Supported Agriculture (CSA), with the mission of producing affordable, nutritious food for their local community in West Dorset and East Devon while also caring for the land.

They started at Lower Hewood Farm (on a two-and-a-half-acre market garden) by providing vegetable boxes to over 50 households, shops and restaurants within a 10-mile radius. As representatives of the Landworkers' Alliance – a grassroots union of farmers, growers and land-based workers around the UK – they campaign for a better food system and the rights of small-scale farmers.

Since 1900, 75 per cent of our plant diversity has been lost. At the same time, the industrialised food system strips nutrients from our soil and from our food. In the UK, austerity measures have also led to a rising inequality over who has access to good food and there has been a dramatic increase in dietary-related ill health. Adam and Dee are trying to address this by growing a wide range of vegetables, including many old varieties such as Tromboncino Squash, which both taste fantastic and have high nutritional value owing to the agroecological farming practices they employ.

In any given season they produce over 200 varieties of 50 different types of vegetable. At the same time, they want to ensure that good food is available to all, not just those who can afford it. Dee says, 'We are farming in a time when there is such inequality in our food system and a stark imbalance over who is able to access nutritious produce and eat well. We want people to feel and know they have a right to good, healthy food and we try and provide it with much care and respect to both the land and our local communities.'

The CSA model enables Dee and Adam to create an alternative to the supermarket which favours large-scale intensive farming. They sell their produce directly to customers who commit to the scheme for the season and come to the farm to pick up their boxes. For Adam, this is one of the rewarding parts of the week. He says, 'One of my favourite sights is on a Thursday afternoon when all the tables in our packing shed are heaving with veg boxes, full of ripe and juicy vegetables, ready to make their way out to the cooking pots and kitchen tables of all our customers.'

Photographer
Sian Davey

FOOD ON YOUR PLATE
FAIR PRICE

SOUTH CENTRAL LOS ANGELES, USA

'Liquor stores, fast food, vacant lots.' This is how Ron Finley, aka the 'Gangsta Gardener', describes South Central Los Angeles, where he was born and raised. He also calls it a 'food prison', where you can buy alcohol and a hamburger on any street, but it is almost impossible to find a piece of fresh fruit. 'I saw dialysis centres popping up like Starbucks,' says Ron.

Upset by the deteriorating health of his community, Ron realised that food was both the problem and the solution and decided to take a stand. One day, in an act of creative defiance, Ron dug up the grass verge outside his house and planted fruit trees and vegetables there instead.

At first, the city council reacted negatively, telling him to remove the garden. This didn't go down well with the growing number of residents and friends who saw the potential of Ron's first garden. After a concerted campaign and critical media attention, the council backed down, even offering its support to the initiative.

Today, the verge, 150 x 10 ft, looks every bit the 'food forest' Ron hoped it would. Banana plants tower over peach, plum and pomegranate trees that provide shade for potatoes, carrots and herbs. People walking along the street take fresh food as they go by, and that is how it should be, says Ron, who is spreading the word and is inspiring others to do the same.

As well as providing healthy, nutritious local food, these gardens are a tool for the education of local people, especially youth, and the transformation of neighbourhoods. 'Gardening is a therapeutic and defiant act,' says Ron. 'When you plant a garden, you're not just feeding people, you're healing the planet, the soil, and yourself, all at the same time.' One of Ron's ultimate aims is to shift the narrative around food growing and make gardening cool. 'If you're not a gardener, you ain't a gangsta,' he says. 'Let your shovel be your weapon of choice.'

Photographer
Stefan Ruiz

DAMN
It feels good to be a
GANGSTA
GARDENER

ROSWITHA HUBER, AUSTRIA, WITH TSHERING WANGMO AND TASHI DORJI, BHUTAN

Roswitha Huber is passionate about good bread. She has made her own sourdough from a blend of alpine rye, grown by her husband and his family, since she first moved to the Rauris Valley in the Austrian Alps as a young bride. This sourdough, made from just three ingredients – rye, water and salt – has a long tradition in the area and is a particular staple for the farmers out in the mountains all day.

Roswitha nurtures the sourdough starter cultures over three to four days, feeding them continuously, until they bubble. 'It is like a son or daughter,' says Roswitha. 'It starts to live and you need to look after it very carefully. Sourdoughs all smell different depending on which environment they're made in.'

From her home Roswitha also teaches others to bake, passing on her love of good, simple food. 'I am convinced,' she says, 'that for the self-confidence of a child, it is essential they have the feeling "I can feed myself".'

Today, Roswitha's 'school in the mountains' has expanded to be part of an incredible exchange programme with farmers from Bhutan, who, despite living nearly 7,000 miles away, grow their produce in a very similar climate and landscape. Like Roswitha, they too work on small-scale farms that produce grains and are looking for extra avenues of income by processing a variety of produce on the farm.

As part of the Organic Farmers Exchange Programme organised by the Bhutan Network, Tshering Wangmo and Tashi Dorji (photographed here) spent two weeks living with Roswitha. Tshering wanted to learn how she could make use of buckwheat for bread-baking, but also learn other practical skills such as milk-processing and making different types of cottage cheese. She also learned about herb cultivation and making fruit jams.

The Bhutanese farmers return to their Kingdom with increased self-confidence and new ways to produce and market their indigenous crops, ensuring that they can continue to grow and sell them to support their families.

Photographer
Zalmaï

MASUMOTO FAMILY FARM, FRESNO, CALIFORNIA, USA

Mas Masumoto inherited an 81-acre organic peach farm in California that his parents had planted nearly 50 years ago, and for many years he grew some of the juiciest and most delicious peach varieties in the region. During the 1980s, as the supermarkets started looking for bigger produce, Mas's heritage peaches were deemed 'too small' and he faced the reality confronting many farmers today: if he wanted to continue farming as a way of life, he would have to pull out his beloved old trees and re-plant more commercial varieties. Ordering a digger, he sat down to write a letter of lament to the *LA Times*. It began like this:

'The last of my Sun Crest peaches will be dug up this fall.
A bulldozer will crawl in, rip each tree from the Earth and toss it aside;
the sounds of cracking limbs and splitting trunks will echo through my fields.
My orchard will topple easily, gobbled up by the power of the diesel engine
and metal rake and my acceptance of a fact that is unbelievable but true:
No one wants a peach variety with a wonderful taste'.

As soon as the letter was published, Mas was overwhelmed by responses from across California and beyond. People begged him to keep the trees and explore alternative markets, vowing that they would be customers. Taking a leap of faith, he did just that. He cancelled the bulldozers and now his seven heirloom varieties of peaches, as well as nectarines, grapes and raisins are in high demand at local farmers' markets and from restaurants sourcing their foods responsibly across the state.

Meanwhile, Mas's 31-year-old daughter, Nikiko, has come home to farm by his side. A Berkeley graduate, Nikiko decided that the most anarchic career move she could make was to take over the legacy of the family farm from her 'best friend', her father Mas.

Today, Nikiko works alongside her father and grandparents, benefiting from two generations of knowledge of how to grow and care for some of the best-tasting peaches in California. Invigorating the family farm with youthful energy, she has recently started an 'ugly' fruit programme called *Organic, Ugly & Fabulous*!, finding homes at discount prices for the slightly odd-shaped fruit that the mainstream supermarkets reject and waste every day. Nikiko says: 'I sometimes wonder why I am doing this... The answer always comes back to family and the belief of feeding people and taking care of the earth.'

Photographer
Carolyn Drake

Brandt

MASUMOTO

Small-scale farmers work together with other species to protect our ecosystems and produce food for our communities.

IN HARMONY WITH OTHER SPECIES

The industrial food system destroys pollinators, insects and the microbial life of the soils through the use of toxic chemical pesticides and fertilisers. A recent study links glyphosate – the most used agricultural chemical ever – with the global decline of bees. Today, the substantial loss of bees and butterflies is said to be directly responsible for a 5-8 per cent loss in agricultural production globally. At the same time, intensive farming practices keep animals in cramped, cruel and often dirty conditions that require the regular use of drugs to control disease; a practice which has led to antibiotic resistance in humans and animals. Furthermore, the appalling conditions in factory farms are breeding grounds for disease, raising the risk of pathogens jumping from animals to human populations.

In direct contrast, small-scale agroecological farming practices work in harmony with other species. They give animals space to roam, they protect pollinators, including an estimated 20,000 species of bees and other insects, birds and bats globally. Farmers understand that they are participating with others in producing food, from the earthworms that digest organic matter to the animals that fertilise the soils. Other species are seen as allies to work with, rather than pests to control and eliminate. The wide variety of crop and animal diversity on small farms also increases the natural biodiversity on the land.

By placing holistic ecological thinking at the heart of food production, regenerative farming systems work to maintain the health of the wider ecosystem on which our food system depends. In the context of climate change, this is the foundation of resilience.

THE PERABÒ FAMILY, FAEDIS, NORTHERN ITALY

Every morning at 3am, Rino Perabò rises in the farmhouse that his family built over 300 years ago, walks out past the portraits of previous generations of Perabò farmers and into the darkness of the barn, where his beloved herd of Pezzate Rosa cows are resting.

Rino knows every one of his 30 cows by name. And once Stella, Colomba, Roma, Parigine, Viola and their friends have been milked, they leave the barn and wander freely around the farm. Everything the cows eat springs up naturally on the farm or is grown by Rino himself. Everything the cows leave behind on their wanderings Rino keeps to fertilise his crops. By mid-morning, the local cheese-producer arrives to collect the 200 litres of milk these cows produce each day and takes it to the nearby town of Cividale for processing.

Rino's small-scale approach to dairy farming stands in stark contrast to 'Big Agriculture's' methods, which see cows confined to cramped sheds all their lives and injected with hormones and antibiotics to ensure maximum milk production. Industrial-scale operations are run as units of production with little consideration for the animals' welfare or the environmental damage they cause. Combined, the world's top five meat and dairy corporations are now responsible for more annual greenhouse gas emissions than Exxon, Shell or BP.

These corporations are stacking the odds against Rino and farmers like him. Operating on a global scale, they undermine the economic life of small dairy farms everywhere, forcing them to sell their cows and close up milking sheds forever. Since 2002, more than half of all Britain's dairy farms have closed. The situation in Italy is very similar.

But despite tremendous economic pressure, Rino continues to nurture his herd and his family's legacy. At a time when farming traditions and rural culture are being lost, he hopes to be an example of an 'old-fashioned farmer'. He wants to show younger generations that this does not mean being backward or irrelevant, but living and working in an ecologically and socially conscious way that nurtures the life of the whole farm.

Photographer
Davide Degano

PASSONE ODDONE
UDINE

NATURALEZA VIVA, GUADALUPE NORTE, ARGENTINA

Now in their 60s and 70s, Irmina Kleiner and Remo Venica look like unlikely revolutionaries, but this pair of now-doting-grandparents spent over 10 years on the run, living in the jungle and existing on foraged food, after speaking out about the rights of peasants in a dictator-led Argentina. Two of their six children were born on the floor of a hastily constructed mud hut before being handed over to friends, while they fled to Europe to await a change in government.

Today, back home and surrounded by their extended family, they run a tree-clad, 120-hectare mixed, agroecological farm in Argentina's North Eastern province – an area more commonly known for its millions of hectares of GM soya beans. Argentina is the world's biggest exporter of GM soy, and the monocrop, which is routinely sprayed with toxic pesticides, dominates the landscape. It is now blamed for many of the health and social problems in the area. One third of the families here have a relative with cancer compared to 3 per cent in the rest of the country.

In contrast, Remo and Irmina's farm, Naturaleza Viva, follows biodynamic principles of agriculture, sometimes referred to as 'super-organic', which emphasises the need to look after the soil with manures and composts, as well as farm a diversity of crops and animals following the laws of nature. Today, the farm produces everything from rice and wheat to a great variety of cheese and dairy products which are sold in the local markets. Remo's passion for the preservation of seeds has seen him grow 15 varieties of rice over three years in order to revive varieties that are no longer available on the market.

Today, the farm is run by a cooperative of 15 families, including Remo and Irmina's children and their 11 grandchildren, who all live on the farm. Through years of hardship and struggle, Remo and Irmina held firmly to a belief that the only way to move society forward and address such pressing issues as poverty, pollution, deforestation and the domination of large corporations was to reconnect people to the land. Naturaleza Viva is an agroecological vision that took many years to realise, and its success shows what is possible when we stop trying to control the natural world. As Remo says, 'If you work with nature, she will work with you. If you work against nature, she will work against you.'

Photographer
Jordi Ruiz Cirera

ZUZANA PASTORKOVÁ, DLHA NAD VAHOM, SLOVAKIA

Along the River Váh, an hour from Slovakia's capital city Bratislava, lies the small village of Dlha Nad Vahom. Zuzana Pastorková spent her childhood summers here, staying in her grandmother's communist-style bungalow, before leaving to work on luxury yachts around Europe. When Zuzana came home seven years ago, she wanted to grow food that carried the flavour and traditions of the vegetables she remembered from her youth and she set about creating a market garden and running a Community Supported Agriculture (CSA) from the seeds and cuttings given to her by the local community.

Zuzana now cultivates 40 different varieties of vegetable, 10 herbs and many different types of fruit tree. When harvested, they are packed into boxes and feed around 40 families in the region. Of this food, 70 per cent is grown from seed that has either been handed down to her by her family or that she has sourced from her travels. Her quarter-of-an-acre garden now boasts beans from Ireland and Hungary, onions from Romania and pumpkins from Cyprus.

All the seeds come with their own story, but her favourite is a strawberry, whose aroma dominates the garden. 'We call it the Budapest Strawberry,' she says, 'because it was brought to the village at the beginning of the last century by a traveller from the village who found it in a garden of an aristocrat family in Budapest. It was then handed from garden to garden. Its aroma is there the minute I open the gate in the morning and it is a strawberry with the richest flavour I've ever tasted!'

The garden's secret heroes, however, are its six Indian Runner Ducks who keep the slugs and snails down by feasting on them. Zuzana calls them her 'girls and boys' and regularly tells them off when they move from eating slugs to eating her vegetables instead. Zuzana puts the success of the garden down to the way everything works together. 'This is very different to industrial agriculture. It's about being in tune with nature, learning from her and being part of her. The magic and mystery are so important. To understand what's happening in nature you have to be quiet and observe and then the answers emerge on their own.'

Photographer
Tina Hillier

YANGDONG, GUIZHOU, CHINA

In the mountains of Guizhou province, Southwestern China, ethnic Dong farmers of the Yangdong Rice Cooperative harvest their rice on 600-year-old terraces. They use farming methods that hail back to the Han dynasty and involve an ancient and symbiotic relationship between man, animals and nature.

Each rice paddy hosts hundreds of species of animals, insects, amphibians, fish and wild plants. From this ecosystem the communities reap a triple harvest of rice, ducks and fish. Over 5,000 farmers from the cooperative produce rice in this traditional way, providing for their families and the local towns, as well as supplying markets in the cities of Shanghai, Shenzhen and Guangzhou.

These ancient methods are in stark contrast with China's agricultural machine which has already surpassed all other countries in agrochemical production, consumption and pollution. Every year, nearly two million tonnes of pesticides are poured into the Chinese landscape. This creates more pollution than all the country's factories put together and causes significant issues for any life in its way.

The Dong people take a very different approach. All nature is regarded as sacred and having a spirit. During harvest, as a way of giving back to nature, a small portion of the crop is left for the birds and even the rats. The trees are considered to be the ancestors of the village: it is forbidden to cut them down and the forests are left untouched. The village has renounced chemicals and machines completely, preferring instead to rely on the 'cow-duck-fish' trinity to control both the weeds and pests. Meanwhile, oxen are revered and considered the driving force of the paddies, as well as the principal fertilisers.

In contrast to many parts of China, where indigenous rice varieties have been replaced by a limited number of hybrid seed, the Dong cultivate 70 varieties that are native to their mountains. Having been nurtured to evolve in this landscape, these rice varieties are able to tolerate and flourish in the harsh, cold local conditions.

Traditional wisdom has taught the Dong to maintain a balance between humans and nature. The community says: 'Animals and all living beings have a spirit and can become more intimate with humans. Even the old trees, springs and rivers, all have spirits too. We must trust them and work with them, not against them.'

Photographer
Zhang Kechun

ZAZA IVANIDZE, MUSHKI, GEORGIA

When Zaza Ivanidze was a boy he used to climb the walnut tree in his back garden to spy on the old lady – Tuta – who kept bees next door. He was fascinated to watch her carry out this centuries-old tradition, in a region famed globally for its honey. Eventually, Zaza was noticed and invited to become Tuta's pupil before going on later to study at the Tbilisi Beekeepers Association in the Georgian capital. These days, he passes his skills and knowledge onto other children in the village, as well as former prisoners who come for lessons, so they too can continue the tradition, sometimes with the help of his now 93-year-old-neighbour and mentor Tuta.

Although village life has changed rapidly in Georgia in the last 50 years, the people of Mushki still swap produce with their neighbours. Zaza, who now has 100 beehives, regularly trades his honey for wool, milk and cheeses such as Meskheti, Eashyi and Tenili which are difficult to make but highly regarded in the region. There is also a bakery where the bread, made from local Meskhetian wheat grain, is cooked over hot stones each day. Communally, the village employs a shepherd to look after the sheep and take them up into the hills. Although Zaza is limited as to where he can sell his honey, as he only has a bicycle to carry one barrel at a time, customers come to him to fill up their own containers from the barrel on his farm.

Zaza leads a more nomadic life than most of his neighbours, as every summer, with the help of some friends, he loads his hives into two trucks and drives them to the Javakheti mountains, four hours away. The bees stay there on the farm of a local Armenian family for the next two months, feasting on alpine flowers, which makes the honey more flavoursome. The family are good friends of Zaza's and fend off bears and other predators, ensuring the hives are well looked after. In spring, Zaza returns the favour by looking after their beehives in Mushki when the climate is better for the bees in the lower regions.

Zaza loves his way of life, which is both his profession and his passion. He has just become engaged and is now looking forward to teaching his new wife how to take care of the bees when she moves to his village later this year. 'I believe that success comes when you love your work,' he says.

Photographer
Antoine Bruy

Small-scale farming puts the community at the heart of its food production, working together to nourish families and enhance community life.

REVIVING COMMUNITY WELLBEING

In 2010, the world passed a threshold, with more people now living in cities than in rural areas. For many, this global shift has meant the breakdown of deeply rooted communities and a radical disconnection from the origins of the food we eat every day. One of the reasons for this shift is the increasing mechanisation of industrial farming and the concentration of land into the hands of fewer and fewer private companies and landowners.

By comparison, small-scale farming is intensely rooted in place. It brings together a community that goes beyond humans, to the wider ecosystems that sustain our food production, and it fosters resilience within these communities. On these farms, local people – from the family and wider community – work together to produce food for themselves and for others. Supply chains are short, work is communal and food is seasonal. This approach helps maintain and revive a pride in rural skills, ways of life and biodiversity, as well as giving young people a viable future on the land.

These communities also encompass urban food-growing projects where those who may otherwise feel isolated in big cities, can come together to work in a meaningful way. Examples are Community Supported Agricultural schemes (CSAs) and farmers' markets, where people buy their food directly from farmers, thereby reconnecting with the source of their food. Supporting small-scale farming like this also rebuilds social cohesion, trust and participation in community life that is increasingly missing in our modern lives.

SANTA TERESA, LA CONVENCIÓN, PERU

In the shadow of the citadel of Machu Picchu lies the farming community of Santa Teresa. For centuries, the land here was colonized by the Spanish, and the Quechua majority were forced to work for a single landlord who chose when to hand out work on their monocrop plantations. This rich, fertile, land was used to grow sugar cane to make liquor, which was handed out freely to keep workers subdued.

As the workers in Peru started to get organised into unions in the sixties, Santa Teresa's inhabitants saw their opportunity to reclaim their sovereignty. They led the way to what became the most sweeping land reform in Peru's history, eventually securing hundreds of small plots for 300 families who now work together to operate a coffee cooperative called Huadquiña.

Today, Huadquiña is at the forefront of a new agrarian revolution across Peru which uses traditional and agroecological methods to enhance biodiversity and respect for nature, as well as the wellbeing of the community. The cooperative also embraces agroforestry techniques, diversifying the varieties of coffee they cultivate, the trees that shade them and the other plants that grow among them.

Huadquiña manager, Hebert Quispe Palomino, says: 'The members of the co-op have a spiritual relationship with their land that dates back millennia. Before we do anything with our land we first ask permission of the Apus (the spirit of the mountains). Our elders taught us that the Apus will tell us when there will be a good harvest. In the post-harvest fermentation, we pay particular attention to the lunar cycle, because the coffee ferments best when the moon is full.'

Huadquiña is also strengthening its social and inter-generational sustainability by employing the community's youth in different roles within the co-op. They see that by drawing on the experiences of young people from the community, they can ensure that they are able to stay and make a dignified living in an age where rural communities the world over face the challenge of young people leaving the countryside.

You can taste Huadquiña's rich and chocolatey coffee in the UK as Cafédirect's flagship Machu Picchu single-origin coffee.

Photographer
Niall O'Brien

EL CHORO, COCHABAMBA, BOLIVIA

Four thousand metres high in the Andean mountains, the small community of El Choro is returning to a system of farming and food production that looked after it for thousands of years. The system is based on the ancestral philosophy of *sumak kawsay*, which permeated indigenous Quechuan life for thousands of years and promotes the coexistence of all living entities.

Sumak kawsay taught generation after generation to live in harmony with themselves, with their community and above all, with Mother Nature. It is now, once again, at the centre of all community decision-making and helping to bring the families back to a way of life that promotes a diverse and healthy diet as well as financial independence.

As part of this way of life, El Choro works communally to take care of its lands. Here, private property does not exist, as it does in the West. Each family has their own plot, which they spend time cultivating, but they mainly work in the communal fields to provide for everyone. Communally, they have restored 150 ancestral varieties of potatoes as well as quinoa and other grains. They have also brought back traditional medicines, started beekeeping, breeding fish and even cultivating fruit trees high up in their mountains.

Another of the central pillars of *sumak kawsay* philosophy is *ayni* or 'reciprocity'. Through this, it is understood that you must give if you wish to receive. So the community works together to look after nature or *Pachamama* (Mother Earth), and in turn Mother Nature takes care of them. This philosophy has now been embedded in the constitutions of Bolivia and Ecuador. The people of El Choro believe that everything in life is interconnected. They say: 'Everything that the individual does has direct and indirect consequences for all living beings. People everywhere must work together to change the models of modern societies to healthier ones. Not only for humans, but for our Mother Nature, *Pachamama*.'

Photographer
Nick Ballon

ALICE HOLDEN, GROWING COMMUNITIES DAGENHAM FARM, LONDON

Alice Holden is the Head Grower at Growing Communities Dagenham Farm, an organic agricultural project that has transformed an ex-council nursery into a productive market garden. The farm supplies vegetables to the Growing Communities box scheme which sends produce to around 1,000 city homes every week.

The award-winning scheme, based in Hackney, East London, is London's oldest vegetable-box scheme and offers a reliable trade model for ecologically friendly farmers, as they guarantee a fair market price for their produce. On such a small site – fewer than two acres – Alice has focused on producing repeat-cropping, perishable and nutrient-rich vegetables such as salads and leafy greens. These are all fast-growing crops that make the best use of smaller urban spaces where land is often at a premium and can complement other foods that require more space to produce.

For Alice, the enjoyment of her job comes from much more than producing the food. In a city where people can feel isolated, the farm is a place that brings people together. 'We have lots of help from volunteers who come from all sorts of different backgrounds. The growing means we all work towards a common goal. Everybody can help in some way. Everyone is needed. Being outdoors doing physical work and making a tangible transformation can take people out of their heads and make them feel better, physically and mentally. The mental-health benefit is a large part of what has drawn me to this kind of work and I know lots of our volunteers feel the same,' says Alice.

With half the world's population now living in cities, the accessibility of urban farms enables city-dwellers to directly experience the connection between food and nature and how this can sustain us in a variety of ways. 'Unless you come into contact with the process of growing food and its connection to the environment why should you care about it?' says Alice. 'Children who come to the farm can go from not having tasted or seen certain vegetables to knowing how to grow, pick and cook them. They are willing to try things they otherwise would not.'

It's a different type of learning from sitting in a classroom, but in our increasingly urban and desk-bound world, it could be one of the most important lessons they receive. Alice is providing a vital way for children to eat healthy food and connect with the fundamentals of life.

Photographer
Rankin

CHAGFOOD, DEVON, UK

Ed Hamer represents a new generation of young farmers who are determined to put farming and food production back at the heart of community life in the UK. With his family, a few growers and an army of volunteers, he runs a Community Supported Agriculture scheme (CSA) on five acres just outside the idyllic Devon village of Chagford. Now in its tenth successful year, the CSA provides organic vegetables and an increasing array of other produce for 100 families living in the area. The community comes together regularly to help harvest and celebrate the agricultural calendar in a traditional way – something that has been lost for most villages around the UK today.

Ed grew up here and knows the landscape intimately. For many years, however, his dreams of farming had to be put on hold as local property prices escalated and local farmland was sold to urbanites for holiday homes. Ed believes land is a basic human right, and the inability of local people to access it fired his determination to re-engage people with agriculture. The CSA actively encourages people to visit and see how and where their food is produced.

'50 to 60 years ago most people would have known the farm or at least the region from which their food came. Now all we have is a faceless network of supermarkets. We wanted to reconnect the community with that cultural aspect of where their food is grown and the skills and knowledge that we have lost. It's about challenging the commodification of food and farming and putting the culture back into agri-culture.'

Today, with kids running freely through the rows of vegetables and even a baby asleep in the harvested potatoes, farm life couldn't look more appealing. Most of the CSA's volunteers are either retirees or new mothers, two groups who often feel isolated. 'There's a real sense of achievement in getting 100 boxes harvested and packed in one day. We all stop for a group lunch and I think most people feel good because it's a genuinely positive and productive thing to be doing together,' says Ed.

IAIN 'TOLLY' TOLHURST, OXFORDSHIRE, ENGLAND

Iain has become something of an international guru in organic agriculture, being called upon to help other farmers around the world when they are struggling and in need of his experienced eye. In recent years he has advised farmers in Mozambique, Iceland, Albania and Kosovo, where he taught communities of women, widowed by the civil war, how to run their farms on their own.

His own operation is a modest seventeen-acre farm within Oxfordshire's Hardwick Estate – the dreamy setting for Kenneth Grahame's book *The Wind in the Willows*. From here he has operated an organic box scheme for 40 years, which now provides over 150 local families with a diverse range of 200 fruits and vegetables. The farm is the epicentre of year-round celebrations that mark the agricultural calendar. Iain feels he has a moral responsibility to the community that goes beyond his customers and ensures that there are always opportunities for people to visit and learn more about food and farming.

A committed vegetarian, Iain made the decision to stop using animal fertilisers some years ago, preferring instead to make his own fertilisers from compost and green manure: a mixture of legumes, clovers, lupins and grasses which improve the structure and ecology of the soil. This also enables the farm to manage its own fertility and not rely on outside inputs: a key component in keeping their carbon footprint to a minimum. Some years the farm has even been carbon-positive, meaning it has sunk more carbon back into the farm than released into the atmosphere – a very tall order in food production.

Iain sees farming and food production from a holistic angle. 'Biodiversity isn't just an extra that you bolt on, it's central to that whole system,' he says. 'As I see it, I manage the biodiversity, and the vegetables and fruit are actually just by-products of that.'

603301

YEO VALLEY, SOMERSET, ENGLAND

The Mead Family have been dairy farming in Somerset, in the southwest of England – within the same 10 miles – for 500 years. Tim and his mother, Mary, run the farm together and oversee the family business which is named after the landscape that surrounds them – Yeo Valley. Now the largest organic dairy producer in the country, the business is still very much a community operation – providing work for 1,700 local people, operating its two award-winning cafés and garden, co-hosting a summer festival every year and critically, supporting other organic dairy farmers in Somerset by buying their milk at fair prices through Omsco – the organic milk suppliers' cooperative they helped create.

The idea of making yogurt started as a solution to use up the skimmed milk left from making the clotted cream they served with their afternoon teas. It was Tim's father, Roger, who saw the opportunity and started selling the home-labelled pots of yogurt from the back of his Morris Minor van. Within a few years this had expanded to supermarkets and they were looking to other local farms for milk. Although Roger had a vision to expand, Tim says, there was never any real plan. 'You just got up in the morning and did what felt right.' Part of doing 'what felt right' was converting the farm back to organic 20 years ago, which was the way all farms operated until chemical fertilisers were introduced post-war.

Tim was brought up on the farm and knows every inch of it from its lake shores to the desolate fields on the Mendip Hills behind. When his father died suddenly 28 years ago, Tim had to step into his shoes and look after both the farm and the business. Mary oversees the breeding and welfare of the British pedigree Friesian cows of which she has an encyclopaedic knowledge. To her delight, these skills were recognised in 2009 when she won the Farmer of the Year Award in the British Food and Farming Awards.

Both Mary and Tim see farming as a custodial duty. 'I don't have any concept of owning stuff,' says Tim. 'This is just what I do. We're looking after this land for the next generation.' A conservation team are employed full time to help with this, planting trees and new hedges to attract wildlife. Twenty-five acres of the farm are given over to providing homes for wild animals. 'Farms can't move, so we're rooted to this community and have to do our best to look after the land and the food we produce,' says Mary. It's this old-fashioned, no-nonsense approach that has kept the family and farm successful and at the heart of the community. As Mary adds, 'We can find alternative sources of energy, but we can't really find an alternative to eating. We all need food.'

AUNTY'S GARDEN, WAIPATU, NEW ZEALAND

On the East coast of New Zealand's north island is a small town called Waipatu. It was once a thriving place with families and gardens, steeped in ancestral heritage. The first Maori Parliament was held here in 1892. Today, many of the families have moved away to find work in the cities. However, thanks to Arohanui (Hanui) Lawrence, also known locally as Aunty, the gardens of the local *marae* – the meeting house – have become not only a food-growing hub, but also a rehabilitation centre and tourist attraction all rolled into one.

Hanui grew up nearby in a large, extended family that came and went, particularly around harvest time. She learned to grow food from her grandparents and her own aunties who planted acres of *kumara* every year – a white sweet potato revered by Maori. The crop fed the family and also provided an income through sales at the gate, a regular custom in New Zealand's rural communities. Hanui remembers the days she and her grandmother would wander into the garden to inspect the 'firstlings' (first fruits of the harvest). The delight of growing food stayed with her and led her eventually to start a garden on an acre of land, adjacent to the *marae*.

Aunty's Garden is now a national treasure – an acre of spiral-shaped vegetable patches interwoven with limestone pathways bursting with the most spectacular colours from every vegetable you can name. The produce is grown for the community and anyone can help themselves. 'We say everything's a dollar but nobody will go away empty-handed because they can't afford it,' says Hanui. The work in the garden is done by a mix of *marae* members and young offenders on community service. 'We get a lot of gang members,' she says. 'They all go home having enjoyed themselves and some even start gardens themselves.'

Marae gardens are now becoming more common in New Zealand as a way of addressing the poor health among Maori people. In Hanui's area, Maori are two and a half times more likely to die from treatable illnesses, many of them linked to poor nutrition. Hanui welcomes everyone, but her favourite time of the year is the *kumara* harvest in April when the family return home to dig up the pink-and-red tubers. Grandchildren and great-grandchildren from the cities work alongside the older generation carrying on a tradition that is over 800 years old. 'My grandmother and I would pick the first fruits of the season and it gave me great joy. I want my grandchildren to feel the same joy that I did'. Like any wise grandmother, Hanui knows that the future of her garden, and her culture, lies in making sure that they do.

Photographer
Russell Kleyn

Many farming and indigenous communities see themselves as belonging to the land, rather than the other way round. As such, they put the protection of their land and its ecosystems above all else.

CUSTODIANS OF THE LAND AND NATURE

Around the planet today there is a frenzied trade in millions of hectares of forests, coastlines and farmland. Thousands of small communities find their ways of life under threat from industries and states seeking to profit by converting their lands and waters into mines, vast monocultures, settlements and more. Despite being faced by violent repression, these communities are standing up to defend their lands and waters and continue as the true custodians of their places.

At the heart of these struggles lies a different way of seeing and relating to nature. For custodian communities, the Earth is our source of life and culture and is, herself, living. This perspective differs greatly from the approach of the industrial-growth ideology, which sees nature as an inert collection of 'raw materials' to be extracted, commodified and sold. The food that these living lands and waters produce is a gift, and this gift confers on us a duty to care for and maintain the health of the ecosystems on which we all rely.

The growing Rights of Nature movement is seeing the convergence of people coming together across the Earth to assert that nature and all her expressions have an intrinsic right to exist, just as humans do. In fact, humans cannot exist without all the other participants in the web of life; while the web of life can exist without us.

THE BORCA FAMILY, BREB VILLAGE, MARAMURES, ROMANIA

In the Carpathian Mountains of northern Romania, the Borca family finishes constructing the last of their 40 haystacks. Made out of Alfalfa and local grasses, this hay will feed their animals for the winter months to come. Every member of the family is involved in this annual event which involves 10 days of mowing fields followed by haystack-making, lasting from five in the morning until nightfall. In Maramures, life still revolves around this crucial ritual. Anuța and Gheorghe Borca, photographed here, had to cut their honeymoon short when they married during haymaking season.

One of the last bastions of European traditional agriculture, Romania has millions of small-scale farms and the highest levels of self-sufficiency in Europe. Over 60 per cent of the country's milk is produced by families with just two or three cows and used by local people within the same village.

However, Romanian agricultural land (referred to as 'black gold') is increasingly sought after by multinational corporations, agribusiness groups and banks that see it as a good investment. Increasingly, small farmers in Romania are at risk of having their homes, culture and livelihoods taken away, as common land is sold off to foreign companies without consultation or compensation. Farmers face becoming landless labourers for the big agribusiness plantations who export their produce. They are having to watch the destruction of their diverse, agricultural ecosystems that have co-evolved here for generations. It is estimated that already around one million hectares (10 per cent of Romanian farmland) is controlled by foreign capital.

In contrast, the Borca family still follows the farming traditions that have been handed down for the last thousand years and which have preserved this landscape for generations. No pesticides are used here, largely because they are not needed and are too expensive to even consider. The meadows of Romania are consequently the most botanically rich fields in Europe. Each square yard contains more than 50 species of flowers and grasses, many of which are extinct or endangered elsewhere. Wild flowers attract bees, other insects and birds, which are on the decline elsewhere in Europe. Without the annual mowing and haymaking ritual this ecosystem would not exist.

People know and understand their landscape, because they have lived in a reciprocal relationship with it for so many generations. Villagers in this part of Romania can name, on average, 120 species of plants and flowers in the meadows that surround them. Anuța says, 'It is our land. We have to take care of it. We have to teach the children the traditions. And teach them something that allows them to survive if they have no job. It's important because the tradition is a treasure. If they learn it, they will be richer.'

Photographer
Rena Effendi

SAN ISIDRO, JALISCO, MEXICO

The village of San Isidro is on the frontline of a David and Goliath story of farming resistance in Mexico. Determined to hold onto their traditional lands, the residents here are sandwiched between giant agroindustrial corporations that illegally took over their land 30 years ago. On one side, the multinational marketing company, Amway, produces freeze-dried food, suspected to be for the American military. While on the other side, the global agrochemical company, Monsanto, grows 'experimental' GM crops.

These companies' white plastic polytunnels stretch for miles and attract workers from all over Mexico, who spend 12–14 hours a day harvesting food for the US market. Some flee because the conditions are intolerable – with overcrowded greenhouses kept at a constant 40 degrees. Others develop health problems as a result of the pesticides they are forced to use. There are families who have left, some publicly denouncing that they were held as slaves for years.

Most of the surrounding towns and villages serve the needs of these workers for the few hours they have off. However, San Isidro has held the line against enormous external pressure. They continue to save their ancestral seeds and grow Milpa – a traditional mixture of crops, including maize, beans, squash, chilli and tomatoes – that has provided a perfect nutritional mix to nourish the soil and local families for generations. No agrochemicals are used here; instead the community use a combination of traditional agricultural methods and new agroecological techniques.

In 2016, a Mexican court confirmed the community's legal right to farm its land. However, with no official body that will dare to enforce the ruling, nothing has changed and the village remains vulnerable to even more land-grabbing. As the young look to move away for a brighter future, the immediate challenge now for the hundred families of San Isidro is keeping their families together. The elders here want to believe their years of struggle have been worth it and they will keep their land for generations to come. But the community of San Isidro isn't just battling corporations but a whole system that sees the natural world as a resource to be plundered for the economic gain of the few.

Photographer
Graciela Iturbide

SELKIE, FINLAND

Located in North Karelia, in the far east of the Finnish Boreal, Selkie and her people belong to an area characterised by dense forests and clear lakes left behind after the last ice age 10,000 years ago. For thousands of years, people here have relied on or supplemented their livelihoods through fishing, hunting and berry-gathering on seasonal rounds taking in marshland, river, lake and forest.

In the midst of this landscape lies Linnunsuo ('Marsh of the Birds' in Finnish), now a big, beautiful wetland where birds stop and nest on their vast migrations to and from Siberia. However, not so long ago, Linnunsuo was an active mine where peat – a potent fossil fuel – was stripped from the surface of the Earth, transported and burned to produce energy, releasing vast quantities of carbon dioxide into the atmosphere.

Motivated both by a need to pay war reparations to Soviet Russia after World War II and the nation's quest to grow and industrialise its economy, by 2000 Finland had ditched or drained over 60 per cent of marshmires like Linnunsuo to make land available for industrial forestry and agriculture, road-building and mining.

But in summer 2011, the death of a large number of fish caused by an acid leak from the Linnunsuo mine led the villagers of Selkie to campaign against the mining company. They became the first community in Finland to get an active peat mine shut down. Since then, starting with the re-flooding of Linnunsuo, they have embarked on an ambitious programme to re-wild their forests, wetlands and the Jukajoki River and protect them into the future.

Local people and their traditional knowledge are playing a central role in this revival, working hand-in-hand with scientists to restore the waterways to health. As well as restoring ecosystems that help cool the planet at a time of climate change, this process is giving people the opportunity to reconnect with nature and the sustenance that the wild provides.

'We need to restore "lost lands" in a new way that embraces traditional knowledge and rural communities, so that we don't repeat the mistakes of the past,' says Tero Mustonen, Head of Selkie Village. 'As the land changes, we need to change with her. When we heal the land, we heal the people.'

Photographer
Joel Karppanen

TATHLINA LAKE, CANADA

Autumn, and three generations of the Chicot family are gathered on the shores of Tathlina Lake in Canada's Northwest Territories. They have come to hunt, fish and gather wild food for the winter months ahead, following in the footsteps of their ancestors from the Ka'a'gee Tu First Nation for whom Tathlina has been sacred for millennia.

These practices have acquired new meaning in light of the resettlements that forced Ka'a'gee Tu People to leave their homes at the lake-edge decades ago, the residential schools imposed on their children and other aggressive colonial policies through which successive Canadian Governments have tried, and failed, to separate Ka'a'gee Tu People from their culture.

It is in the spirit of resisting displacement and cultural loss that the Chicots and others return to Tathlina Lake for days or weeks at a time. For those who once lived on the lakeshore, it is a homecoming. For the younger generations, like Tarek Chicot, it is an opportunity to learn traditional skills, language and spirituality from elders such as his uncle Lloyd and grandfather Gabe Chicot. For all, visits to Tathlina are times to sit, talk, eat and sing together and reconnect with their ancestral place of origin, restoring pride in a culture that settlers have labelled backward, dead or dying.

In this part of Canada, where fresh food must be trucked over 16 hours by road and is only available at very high prices, the meat, fish, berries and other riches that come from Tathlina remain vital to the traditional economy. Because of the lake, people can eat well and stay healthy.

Ka'a'gee Tu People are also adopting new practices in response to challenges such as climate change. For the first time in their history they have started growing food, opening a community garden to supplement their traditional diet and reduce their reliance on expensive imports.

Learning new skills while reaffirming age-old practices, valuing both traditional ecological knowledge and new technologies, the Ka'a'gee Tu say that a person today must 'be strong like two people'.

Photographer
Pat Kane

Tahoe

THE YASSIN FAMILY, ANIN, THE WEST BANK

Every November, after the first rains have fallen, the Yassin family and their friends meet at dawn for the olive harvest. It is back-breaking work in the full heat of the day but by sunset, everyone gathers under a favourite tree to drink sweet mint tea and eat flatbreads with za'atar (wild thyme, sumac and sesame paste). From toddlers to grandparents, all are present for this centuries-old tradition: a time to reunite and share knowledge with younger generations. The whole harvest will take a month to complete, with the olives being taken daily to the Anin oil press, where the family, like thousands of other Palestinian farmers, sell their oil through the Palestinian Fair Trade Association and Canaan Palestine.

This region is one of the centres of origin of agriculture, where farming began, and the olive tree holds a special place. It is revered as a symbol of the Palestinian people, of their heritage and their connection with the land. This relationship is currently jeopardised by ongoing conflict and a ten-foot barbed-wire wall which divides Israel and The West Bank. More than half of the Yassin family's 1,000 olive trees are on the other side of this wall, which they can only access with a permit two days a week.

Today, Hassan and Mohammed Yassin farm five hectares of olive groves which were inherited from their grandparents. They can trace their family back at least four generations and many of their trees are *Rumi* trees, so called because they were planted in Roman times and are around 2,000 years old. On terraces built by their forefathers, they continue the olive tradition, nurturing and cherishing the trees as their heritage to pass onto their children.

As well as the olives, there are chickpeas, anise, wheat and other crops that are cultivated between the olive trees. When the crops are harvested, Mohammed's sheep and goats weed and fertilise the fields. The animal manure and the crop rotations feed the soil, providing nutrients to the soil humus, the microbes and the trees, maintaining a vibrant and healthy ecosystem. It is an ongoing exchange, reciprocity between the land, the animals, the crops and the farmer.

The trees are not only the family's inheritance, they are also the gift to future generations. As long as they grow, this inheritance will be passed on. Mohammed looks after the trees like his own children, watering new ones by hand until they are 16 years old (then the rains will suffice). 'You must feed the tree, just as the tree feeds you. You must love the tree as much as the tree loves you,' he says. When they get older, the trees give back and will, he hopes, look after him and his descendants for another 2,000 years.

Photographer
Tanya Habjouqa

LEIGH KUWANWISIWMA, KYKOTSMOVI, ARIZONA, USA

Leigh Kuwanwisiwma is 68. He still farms his corn, sunflower and squash fields on the Hopi Reservation where he has lived all his life. It is a beautiful but unforgiving landscape that requires a great deal of skill to farm and produce food. To be Hopi means to live with respect and reverence for all of life and to live in accordance with the instructions of Ma'saw, the Guardian of the Earth.

Leigh has dedicated his life to the preservation of these Hopi traditions and as a seed guardian has brought back 40 varieties of indigenous seeds to Hopi lands, in a milestone rematriation of indigenous seed. The most important of all crops are the 17 Hopi corns, especially the three blue corn varieties; *sakwaqa'o*, *huruskwapu* and *maasiqa'o*, which are considered sacred crops given to the Hopi people by Ma'saw to protect and nourish them. The Hopi recognise that these small corn kernels carry the seed of life and so they are grown and looked after like their children. When he was young, Leigh's father told him: 'Never, never let go of the corn. When I pass on, carry on the corn.'

In the deserts of Arizona, where it snows from January to May and the average rainfall is just 10 inches a year, this is not an easy exercise, and an understanding of how to find and preserve natural moisture is essential. The Hopi bury the corn seeds eighteen inches into the ground so they find and maintain moisture to help them grow. Only Hopi corn is able to be grown at such a depth, as it has been cultivated and nurtured by the Hopi people to adapt to this hostile climate. Seeds like these, which have such distinctive characteristics, are invaluable to humankind, as they have adapted to grow in the kind of extreme conditions that climate change is forcing upon us.

Leigh's skills in dryland farming have been handed down through the generations. Everything is done by hand and depends on keen observation of the landscape. As there is no irrigation in the desert, the Hopi rely on the snowmelt from the mountains to flood the fields. The Hopi also hold ceremonies for rain and petition the clouds to come and bless them. Leigh knows the importance of his work. 'One day,' he says 'our corn will become very, very important for human survival.'

Photographer
Jane Hilton

Soils are the basis of life – for plants, animals and humans. The majority of our food comes from the soil.

REGENERATING OUR SOIL

Today a third of our planet's soils are severely degraded and fertile soil is being lost at the rate of 24 billion tonnes a year, according to the UN Food and Agriculture Organisation (UNFAO). Whilst exact figures vary, the UNFAO also states that globally we have just sixty years of harvests left at current rates of soil loss and degradationn. Industrial agricultural practices, such as heavy tilling, large monocrop plantations and the constant use of chemical fertilisers and pesticides, have depleted our soils to the point where they can no longer replenish themselves. Unless we radically change the way we produce food, we are in danger of leaving wastelands to our children.

Around the world, small-scale farmers using agroecological farming techniques build healthy, fertile soils. They practise methods such as no-till agriculture, planting cover crops to reduce soil exposure, grazing animals in a way that mimics nature and rotating their crops to utilise different nutrients in the soil. Rather than using chemical fertilisers, they feed the land with composts and manures.

These farming methods allow rich soil ecosystems to flourish, nurturing the worms and tiny micro-organisms that digest organic material left behind by plants and animals and recycling it to support the production of healthy, nutritious food. By nurturing the soil, the soil is able to nurture us all.

"Soil is the great connector of lives," writes the activist and author Wendell Berry, "the source and destination of all. It is the healer and restorer and resurrector, by which disease passes into health, age into youth, death into life. Without proper care for it we can have no community, because without proper care for it we can have no life."

BASSIERI COMMUNITY, GNAGNA PROVINCE, EASTERN BURKINA FASO

Eight years ago, the women of the Bassieri village faced a food crisis. Drought and eroded soils, degraded through the cutting of trees and over-reliance on monocultures and chemicals, had left them with little food to feed their families. The situation was so desperate that the women were forced to break into the giant termite mounds that dot the arid landscape and steal grains from the insects in order to feed their children.

In the midst of this crisis, 43 women of the village came together to create a Women's Growing Association. Led by Pobadou Lankoande, they worked hard to revive the knowledge of traditional farming practices through learning from the elders of the West Sahel. These ancient traditions have been enhanced by training in agroecological farming practices provided by a local organisation, Association Nourrir Sans Détruire (Association to Nourish Without Destroying).

First they set about improving their soils by reviving water and soil conservation techniques that were once commonplace in traditional agriculture in the region. Small trees were pruned and the regrowth protected, which helped to stabilise the erosion of the soil and collect moisture. Stone barriers were built along the contour line of their plots to minimise water runoff. 'Half moons' – small crescent-shaped basins dug into the soil and filled with organic manure – were placed strategically to catch the run-off. In addition, the women created *zaï* pits – or 'planting pockets' – around 20 cm deep. Each of these pits is filled with three handfuls of organic manure, compost, or dry plant biomass before the seeds are sown.

The key to the success of the Women's Growing Association is collaboration. The women work closely with one another, cultivating both common and individual plots and continuing to learn from the elders. They recognised that diversifying their crops was also critical. They focus on growing a mix of crops, intercropping cereal (millet or sorghum) with leguminous (cowpeas, peanuts, sesame), in order to fix nitrogen into the soil, increasing its fertility.

The women are now producing enough nutritious food not only to feed their own families, but also to sell their surplus vegetables at the local market. Some of the funds from these sales are pooled in a credit co-op and then reinvested in priorities identified by the group.

Fatou Batta from Association Nourrir Sans Détruire proudly points out that 'it is the women who are the rehabilitators of this land'.

Photographer
Andrew Esiebo

ZARRAGA, PANAY ISLAND, PHILIPPINES

Like many smallholder farmers around the world, the rice farmers from Zarraga on Panay Island in the Central Philippines, found themselves trapped in a cycle of using chemical fertilisers to grow their crops. What had been sold to them initially as a way to increase their yields very quickly became a trap, leaving them with soils depleted of nutrients and unable to grow anything without more chemical inputs.

The cycle left these once self-sufficient farmers dependent on agrochemical companies who sold them not just the chemical fertilisers but the rice seed engineered to depend on it. Farmers were forced to sign contracts that obliged them to buy all their inputs from the same company (chemical pesticides, fertilisers and seeds) and over time, this ended up costing them more than they could earn. 'You pay, then you borrow again. The debt never runs out,' says Leticia Subong, photographed here.

For farmers like Joby Arandela, Johnny Subong and Edgar Tono (also photographed) there seemed little way out until they learned about SRI (System of Rice Intensification), a method of rice production which is used in 55 countries from India to Madagascar. SRI is a technique where rice seedlings are planted further apart and given less water, allowing the roots to take in more oxygen and grow better. This simple practice has been known to increase yields by up to four times in India.

Now, after four years, Johnny and his mother, Leticia, are finally out of debt and have become fully organic. The high yields – almost double what they were before – have eliminated any need for chemicals and they are now able to replace the hybrid seeds with indigenous Panay varieties. The chemical fertilisers caused Leticia's nails to turn black and her water buffaloes to develop gangrene on their hooves. Now, native fish like Puyo and Pantat have returned to the paddies and she and her buffalo are thriving.

Today, 40 farmers on the island have formed the ZIDOFA Association and are able to share the abundance they have created by working with the land. They sell direct to their customers and earn far more for their hard work, skipping out the exploitative middlemen. Tests have also shown that in Joby's soil, organic matter has doubled since he began SRI. 'Before, the soil was hard and cracked easily during droughts,' says Leticia. 'Now, the texture has changed, the soil is soft and when you step on it, your feet sink deep into the ground.'

Photographer
Hannah Reyes Morales

YOU CAN'T
eat
WITHOUT FARMERS

COLIN SEIS, WINONA FARM, NEW SOUTH WALES, AUSTRALIA

Colin Seis' farm in New South Wales was all but destroyed 25 years ago in a devastating bushfire. He lost 4,000 sheep, all the buildings and almost his life. The disaster left Colin with no money to buy chemical fertilisers, which the farm had depended on for 30 years.

'My father was a part of the Green Revolution and adopted all the fertilisers and pesticides going. For a while they worked, but we were starting to see things go wrong; the soil was more acidic, we had more weeds and more pests. We didn't know at the time, but ecologically things were beginning to crash.'

Overnight, the Seis family needed a way to farm that didn't cost any money. Colin was the fourth generation on the land and had grown up hearing stories of the extraordinary diversity of the native grasslands that grew in the area. The grasses had evolved to tolerate the local soil conditions and unlike the European or 'improved' grasses that fed the sheep, the native varieties didn't need chemical inputs. Using simple methods, Colin encouraged them to regenerate.

Furthermore, over a few too many pints with a local mate, Colin decided to follow nature's lead and instead of ploughing up the fields – as industrial agriculture dictated – he tried planting cereal crops such as wheat, oats and barley straight into the grassland in the dormant winter months. 'I didn't have any formal agricultural training, so to me, it made logical sense. Why did we need to kill everything to sow a new crop?'

Colin's punt paid off. Not only did the cereal crops flourish but this new system rejuvenated soil health and local ecosystems as well. Kangaroos, koala bears and hundreds of native birds returned to the farm. Colin also experimented with a system of land management known as 'mob grazing', which keeps the animals on the move every few days – much like wild species – allowing the land to recover.

Today, Colin's crazy idea is called 'pasture cropping' and is employed on three million acres around the world. He is also working with aboriginal people and academics to revive native grass seed as a food source. Colin gets angry that farmers are led to believe that they must have chemical fertilisers to be farmers. 'It is just making a lot of companies wealthy.' He is proud to have restored the grassland his great-grandfather moved onto and has made the farm more profitable than it ever was before. 'Being a farmer now is very easy and that's because I just let Mother Nature run it for me.'

Photographer
Katrin Koenning

zero

AGAINST THE GRAIN FARM, ZIONVILLE, NORTH CAROLINA

In 2011, Holly Whitesides and Andy Bryant chose to leave the city and their careers behind them and do something radical. They moved to a 20-acre farm at the foot of the Blue Ridge Mountains in North Carolina, and started to regenerate the land.

Andy and Holly's farm, now called Against the Grain, was a tobacco farm when they took it over seven years ago. In its former life the land had been heavily sprayed with chemical fertilisers which had devastating impacts on the soil and all that grew on it. When renovating the farmhouse, the couple found a soil survey from the 1940s that showed them all they needed to know about what had happened in the intervening years: they were farming on subsoil, all of the topsoil had been eroded.

Andy and Holly set about reviving the soil using a system of agriculture called biodynamics, which has many similarities to organic farming. It emphasises the use of manures and composts, but goes further by integrating these practices with a holistic understanding of the entire ecosystem, mimicking nature and observing its cycles closely. Holly says, 'Biodynamic farming allows us to see the farm as a whole organism with its own path, and its own history and helps us to make decisions with a long view.'

As part of their commitment to rejuvenate the top soil, Holly and Andy made the decision to stop tilling the fields. Ploughing to plant crops is as old as agriculture itself, but many now realise that doing so destroys the microbes in the soil and releases carbon dioxide into the atmosphere. As such, 10 per cent of farmers in the US now practice no-till agriculture.

Today, Against the Grain is home to a diverse and abundant mix of vegetables, cows, pigs, turkeys, chickens – and children – all happily coexisting on the land. Andy and Holly operate a vegetable-box scheme as well as selling from the farmers' market and directly to restaurants. They are part of a new movement of family farmers that are helping to shift North Carolina away from a dependency on cash crops like tobacco to find new agricultural economies that are healthier and regenerative for both the people and the land.

Photographer
Matt Eich

AGAINST the GRAIN

MUFF MASTERS
PERFORMANCE EXHAUST
Kannapolis, NC
ROTELLA T
TRIPLE

Without water there is no life, no food, no food sovereignty. Agroecology takes care of water and its cycles, using a range of ways to protect its purity and flow.

PROTECTING OUR WATER

Oceans, rivers, rains, lakes and aquifers are crucial for feeding the world. We need them to water crops and provide drinking water for our animals and ourselves. According to a 2011 report, agriculture represents over 70 per cent of water consumption around the world. Yet chemical run-off from industrial agriculture is one of the biggest polluters of freshwater systems globally and the agroindustry is a major contributor to climate change, a leading cause of desertification and ocean acidification.

Small-scale fisher-people and farmers know that water is precious. Over the millennia they have developed locally adapted ways of fishing and farming that keeps water clean, healthy and flowing. There are 800 million small-scale fisher-people in the world today who use methods handed down through generations, which respect the life cycles of fish so that their populations remain steady and protects the waters in which they live. Agroecological farming regenerates the wider ecosystem – the rivers, wetlands, forests, springs and mangroves – increasing the capacity of the land to deal with drought or flooding.

These communities recognise water as the source of life and they fight to protect their rivers, lakes and oceans from the abuses of the industrial system and from a growing water-grabbing phenomenon, as fresh water becomes ever more scarce.

KALIX, KUSTRINGEN, SWEDEN

Joakim Bostrom and his friends have fished the local inlets around their villages in Kustringen, northern Sweden since they were boys. As well as a source of food and a means of income, fishing is a way of life here. It is woven deep into the language and knowledge of local people.

In the local dialect, called Kalix, Joakim's grandfather taught him the signs and symbols that can be seen in nature that indicate the best locations to fish, the best methods for fishing and when to go out with the nets. But his generation could be the last to pass this knowledge on to their children.

New Swedish legislation banning all fishing in waters less than 3 metres in depth threatens to criminalise the ancestral fishing methods of Kustringen's fishermen. In addition, new European Union legislation criminalises all selling of fish, except by those few individuals who have a commercial fishing licence. These bans would not only abruptly take away the Kustringen fishermen's source of food and income, it could alienate them from their culture and the seascape that has nurtured it.

'Today it is impossible for me to hand over this knowledge and teach future generations the important knowledge connected to fishing. How do I explain to my child that it is illegal to fish? How shall I respond to my children when they want to put their nets down where my grandfather taught me to fish?' asks Joakim.

The new legislation is designed to protect endangered species, such as Havsöring or sea trout. However, the blanket bans negotiated far from Europe's Arctic North, don't take into account the many plentiful species, such as pike, perch and burbot, that also inhabit these shallow waters and have been fished by local people for centuries. The fishermen of Kustringen want the authorities to recognise their ability to steward nature and maintain the health of the fish, based on indigenous knowledge.

The future of their children's heritage now lies in making the invisible visible. 'It's not easy to put into words the difference between scientific knowledge and our own local knowledge,' says village elder Peder Nilsson. But translating Kalix, which holds the depth of the fisherman's understanding and connection to place, into a language that government officials might understand, may be the only way to protect this traditional way of life.'

Photographer
Clare Benson

REPORTAGE
man får till hästarna är

CAJAMARCA, TOLIMA, COLOMBIA

Bernain Vargas Fandino and his family are proud third- and fourth-generation arracacha-growers in Cajamarca, Colombia. Arracacha is a unique Andean tuber – somewhere between a carrot and celeriac root – and the small town of Cajamarca is its capital, producing more than anywhere else in the world, an average of 100 tonnes per day.

In recent years, Cajamarca has reached international prominence for another reason. In March 2017, the people of Cajamarca held a popular consultation and voted overwhelmingly to block plans to open a mega gold mine. Known as La Colosa, the mine would have generated 100 million tonnes of contaminated waste rock and destroyed huge parts of the surrounding landscape, threatening the high Andean wetland moors that provide most of Colombia's fresh water.

The vote was organised collaboratively by local farmers, environmental activists and a youth collective who led a decade-long campaign against what would have been one of the largest gold mines in the world. With victory secured, focus turned back to the arracacha: heralded the 'true gold' of the region.

Ethical Colombian restaurant chain Crepes & Waffles wanted to support the local farmers and set about designing a new menu with the arracacha at its heart. Since October 2017, the company has bought two tonnes of Arracacha every month at triple the market price directly from Bernain's farmers' association, ASPROSSAN. They have sold more than 10,000 arracacha dishes and given a facelift to an almost forgotten indigenous vegetable.

The struggle of the farmers in Cajamarca has now been celebrated in paintings and exhibitions across Colombia, which makes Bernain very proud. For him, the arracacha is about far more than a food source. It represents the connection between the people of Cajamarca and their land, which they can now pass on to future generations – to continue protecting their land and this ancestral crop.

Photographer
Federico Pardo

ALOHA

SHANNON ELDREDGE, CHATHAM, MASSACHUSETTS

In the early morning mist of Cape Cod, Shannon Eldredge drives her family boat out to Nantucket Sound and sets up the fishing weirs that have been the source of her family's livelihood for three generations. The 45-foot long hickory poles are driven into the sandy sea floor and act like an underwater fence, redirecting the fish into a net where they can be harvested. Squid in the spring and Spanish mackerel and bluefish in the summer. Unlike the commercial boats that trawl the New England coast, traditional weir fishing is small-scale and respects the breeding season of the fish, ensuring that the diversity of the types of fish and their populations have remained stable over generations.

For 35-year-old Shannon it is also a way of life, although her family never intended her to take it on. After a childhood of playing around the docks, her parents were determined that she should move away and get an education. Cheap imports of fish were undermining local markets and fish stocks were on the decline owing to the devastating impact of industrial-scale fishing operations. To raise money for college, Shannon was given a job clamming – digging shellfish on the sandy flats. But rather than put her off, it drew her further into the family tradition. 'I loved every minute of it,' she says. 'I learned what it means to be self-employed and how to have a strong work ethic. I did go to college and never missed a class because I knew it was going to cost me 200 lbs of clams next season!'

After Shannon graduated, she returned to work alongside her father and uncle. The first season was a disaster: 'The fish just never swam into the nets... all that work for almost nothing. It was devastating,' she says. Shannon decided to try one more year and set up the weirs again. This time, with the 'hail coming down so hard, the water was frothing jade green and snow white.' However, that day they had a haul of 9,000 lbs of mackerel – 'more fish in one day than we'd had the entire previous season!' says Shannon.

These days, Shannon hedges her bets and finds other ways to supplement the family income. 'We piece it together because we love working on the water, and feeding people.' Despite a college education, Shannon says she enjoys physical labour best. 'Most people pay crazy amounts of money to exercise. I get paid to build muscle and sweat! Also, I learned a special set of skills directly from my father. That's rare in our culture these days... and I'm proud to take on the gift of his knowledge.'

Photographer
Holly Lynton

XOLOBENI, SOUTH AFRICA

Above the craggy cliffs, crashing waves and white-sand beaches of South Africa's Wild Coast, the indigenous Amadiba People of Xolobeni live a rich subsistence life. In their ancestral homeland, they fish and farm on small plots, raising chickens, growing maize, sweet potatoes and other vegetables that provide food for their families: critical in an area two hours from the nearest shop.

In recent times, however, Xolobeni has attracted new admirers. The mineral-rich sands of the Wild Coast have brought mining companies to the region, among them Minerals Resources Limited (MRC), an Australian mining company with British investment. MRC plans to gouge out dune sand in an area 22 km long by 1.5 km wide, as part of a project that threatens the Amadiba People's ancient lands and waters with pollution and sterility.

The project would cut the villages of Xolobeni off from the sea, displace over 1,000 people from their homes and require ancestors to be dug from their graves and moved. Many Xolobeni residents perceive the project as an existential threat to their culture, spirituality and ways of living. An overwhelming majority have opposed MRC's project from the start and their resistance has been steadfast and resolute.

Getting organised as the Amadiba Crisis Committee (ACC), the community have written petitions, protested on the streets of South African cities and created blockades along the coastline. This resistance has been met with deadly violence. In addition to the beatings, death threats and abuse experienced by many local people, four vocal opponents of MRC's mine have been killed in recent years. The latest was Sikosiphi 'Bazooka' Rhadebe, Chairman of the ACC, who was murdered in his own home. Two years on, his killers are yet to be brought to justice.

To date, the people of Xolobeni have successfully stopped MRC's mining operations and are now at the forefront of a campaign uniting communities across Southern Africa to assert their Right to Say No to unwanted mining.

But for leaders like Nonhle Mbuthuma, a spokesperson for the ACC, the price of defending land and life is high. She is forced to live away from the community and under armed guard for fear of her life. 'I wake up each morning and thank God I am still alive,' she says. 'My husband and my family and friends are worried. They tell me to go into hiding. But I can't do that. We want to keep our land and keep our lives and keep the way we have been living for generations. We don't want to lose our land.'

Photographer
Lindeka Qampi

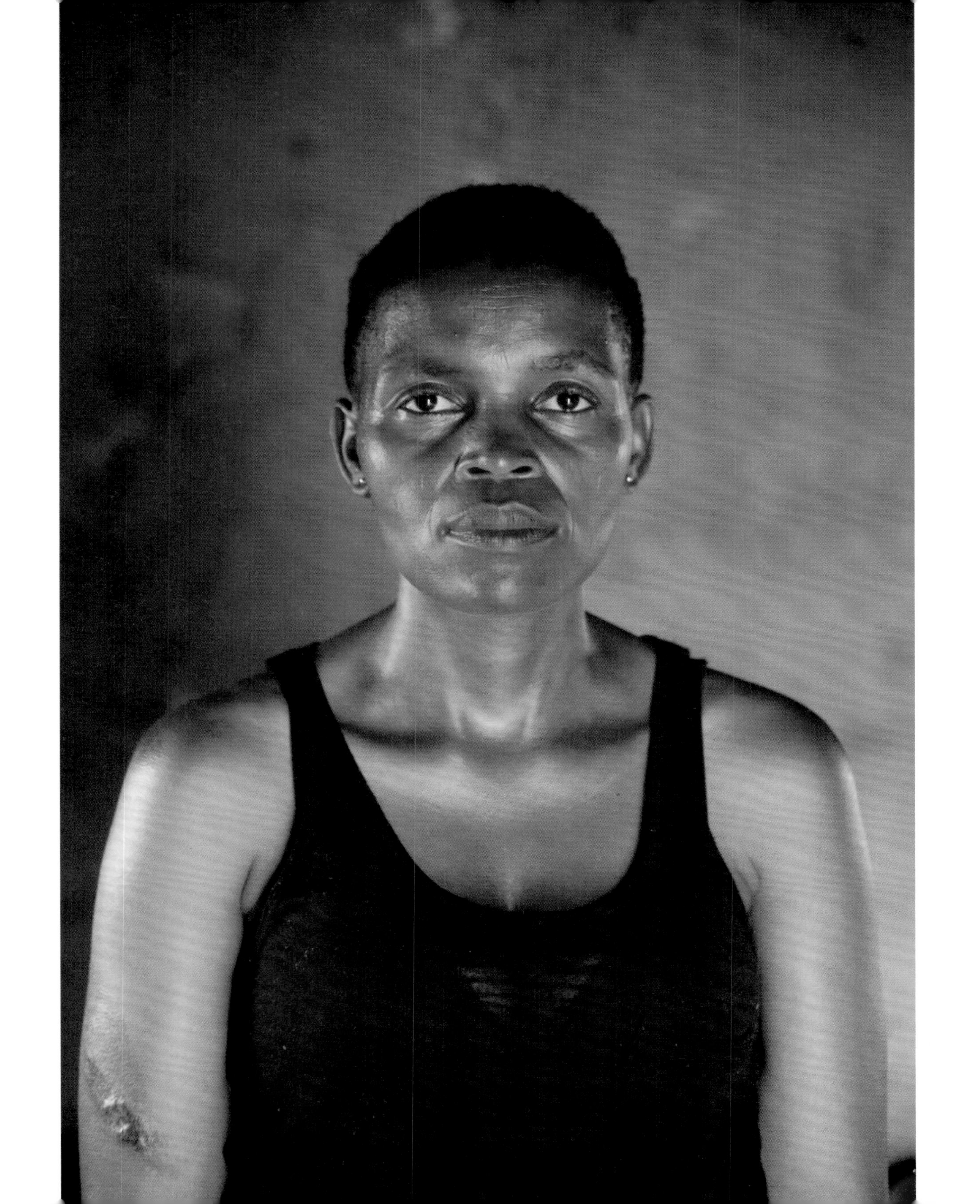

Farming, hunting, fishing and gathering still provide livelihoods for the majority of the world's rural people. These practices are ways of life that bind people together in a common culture.

PROVIDING MEANINGFUL WORK FOR ALL

Farming, hunting, fishing and gathering still provide livelihoods for the majority of the world's rural people. These practices are ways of life that bind people together in a common culture.

In parts of the world where agriculture has expanded and become increasingly mechanised, this is no longer the case. People have become dispossessed of their lands and livelihoods and forced to work on plantations or in factories, which leave them exposed to health risks from machinery and pesticides. While governments subsidise the industrial food system, the workers subsidise their wages with charitable handouts – more than half of US fast-food workers rely on food stamps. In many parts of the world, the working practices of the industrial food system violate Human Rights, including countless cases of slavery and child labour. Sixty percent of child labour around the world is in industrial agriculture.

Over the last three decades, small-scale farming, fishing and hunting communities across the planet have been getting organised, growing their social movements such as La Via Campesina, to coordinate creative alternatives to the dominant industrial-growth economy. Together they have developed the seven pillars of Food Sovereignty, focusing on regenerating land, water systems, community, economy, and sovereignty from the bottom up. This includes a diverse range of pathways such as 'Community Supported Agriculture' (CSA), which directly reconnect local people with farmers in a mutually enhancing relationship; and the Transition Town Movement, which encourages community initiatives to produce as much of the food and energy required locally.

Small-scale farming and fishing communities create meaningful livelihoods for more people in diversified economies than the globalised industrial agricultural system, because there are more opportunities for people to regain skills, creativity and autonomy.

FILHOS DA TERRA, ALAGOAS, BRAZIL

Forty years ago, the families of the Filhos da Terra community worked as labourers on the sugar plantation of Ouricuri, earning minimal wages to produce sugar for the insatiable export market. When the landowner refused to pay their wages, they were left with two options. They could either abandon their homes and migrate to the overcrowded slums of nearby cities or take back the very thing they needed to lead a more dignified and fulfilling life – land.

The community occupied unused plantation land and started to build shelters and grow food. The landlord hired armed militia to remove them. The next 13 years were a struggle of violent conflict until finally the government granted the community the right to stay and feed themselves from the land they grew up on. Today, Filhos da Terra is just one of thousands of communities across Brazil that form the biggest social movement in Latin America, MST (Movement of Landless Workers).

In Brazil, 85 per cent of all agricultural land is used for the production of commodities for export, such as soya bean, corn, beef and sugar cane, while at least four million landless workers struggle to feed their families. Since 1984, MST has helped 400,000 families to seize unused land, obtaining the land rights from the government and supporting them to produce healthy organic food. Within the framework of MST, farmers have access to transport, financial aid to scale up their production and markets to sell their produce.

MST also provides a wide range of social programmes. They have set up around 2,000 schools, educating adults and children, as well as running their own university and agroecological training centres. Filhos da Terra now has a healthcare centre, an evening school, an active youth group which fosters small businesses such as beekeeping – run by the teenagers – and a women's cooperative. Community decisions are reached through discussion together and women make up 50 per cent of the coordination positions.

Today, the community of Filhos da Terra produces over 100 tonnes of cassava, as well as honey, beans and other diversified food. From destitute plantation labourers, the people here have become empowered, literate food-producers. MST leader João Pedro Stédile says, 'In a country where the rural areas have always been relegated to material, aesthetic and intellectual poverty, the community is proud to have regained their citizenship and to no longer be just subordinates, which is an essential condition to paving a path towards emancipation.'

Photographer
Bruno Morais

HAPPY

THE CECAQ-11 COOPERATIVE, SÃO TOMÉ ISLAND, AFRICA

São Tomé is a volcanic island, covered in lush, verdant rainforest, 155 miles off the Coast of Gabon. Together with its sister island, Princípe, it is sometimes described as the Galapagos of Africa, as it has more unique species per square mile than anywhere on Earth. Only 30 miles long and 20 miles wide, it offers the perfect conditions for growing cocoa, and during Portuguese rule, which lasted 500 years, it became the largest producer of cocoa in the world.

When São Tomé and Princípe declared independence in the seventies, the land from the old cocoa plantations was divided into smallholdings and distributed among the local people who used it to farm. Eventually, they organised themselves into cooperatives and decided to grow cocoa again. Two of them, CECAB (Cooperativo de Exportação de Cacau Biologico) and CECAQ-11 (Cooperativa de Exportação de Cacau de Qualidade) developed partnerships with Fairtrade companies Divine Chocolate and Cafédirect, which turned their high-quality organic cocoa into premium chocolate products to sell in the UK and beyond. As well as paying the world price for the cocoa, these companies also pay an additional Fairtrade premium which is used to regenerate the farms and communities.

Lary Gaspar, Hortência Pina and Fatima Silva are among the 1,135 members of the flourishing CECAQ-11 cooperative, which is based in the south of the island and run by Adalberto Luis. By joining the cooperative, farmers are guaranteed a regular buyer and payment throughout the year; they also decide collectively how to invest the Fairtrade premium. So far, CECAQ-11 has used the extra funds to build a day-care centre and nursery, to bring electricity to villages and buy a TV for communal use. The next priorities include better access to water and improving the roads to their still remote communities.

A third of the cooperative farmers are women who grow food to feed their families and also earn a regular income from the cocoa they now grow. Hortência is the secretary of the local association of her community, Monte Bello. She has six children and has built a house and bought a motorbike – the transport of choice on the island – as a result of the income she has generated. Meanwhile, Fatima, whose cocoa farm is on the steep slopes of another village, is the leader of her local women's group. Both women feel empowered by taking leadership roles in the farmers' cooperative. Fatima says her biggest aspiration remains 'the health of everyone in the community'.

Cafédirect sells São Tomé hot chocolate and Divine launched an organic dark-chocolate range in 2018, made from the cocoa grown by the farmers of São Tomé.

Photographer
David Chancellor

SÓLHEIMAR, ICELAND

Nestled in a remote Icelandic valley, Sólheimar is the oldest eco village in the world. It is the vision of a pioneering teacher, Sesselia Sigmundsdottir, who dreamed of building a self-sustaining community where people and nature could work and evolve together.

Sigmundsdottir was inspired by the teachings of Austrian philosopher Rudolf Steiner, whose prolific work also led to the foundation of the Waldorf Steiner Schools, biodynamic farming and the creation of the natropathic medical company, Weleda. In 1930, she leased land in the southwest of Iceland, with its own hot spring, and started to build a community of people from different backgrounds, but particularly those who might be marginalised elsewhere in society.

Today Sólheimar has 120 permanent residents, 45 of whom have learning difficulties. Here they farm together, growing a wide variety of crops including tomatoes, cucumbers, potatoes, sweet peppers and herbs in geothermal greenhouses. Thanks to the combined efforts of the community, Sólheimar has become one of Iceland's largest producers of organic vegetables and the home to its only organic forest nursery. It also welcomes 30,000 visitors a year, many of whom stay in the on-site guesthouses.

Algeir Snær Backman, 19, is the son of one of the permanent families in residence. Algeir says caring for the plants in the geothermal greenhouse has given him a different perspective on the food he eats. 'Growing food makes me feel like I'm part of the community and that my work has a positive impact. I consider it a privilege to have been raised here.'

Sólheimar works closely with Iceland's Ministry of Labour to provide opportunities for unemployed people to get vocational training and build their confidence. Senior citizens are also warmly welcomed. Through its focus on food-growing and creating a rich culture of place, in Sólheimar the healing of land and the human community go hand-in-hand.

Photographer
Jack Latham

RINCÓN GRANDE, CHIMALTENANGO, GUATEMALA

The Perén family farms a small, two-hectare plot of land in the village of Rincón Grande, in the highlands west of Guatemala City. Working together in the fields, they harvest their strawberries twice a week, which the mother, Angela, then takes to Guatemala City to sell. To reach the market by opening time, she must get up at 2am and travel on the local bus. She will not return home until 4 to 5pm in the evening.

It is tiring work, but the Peréns have survived many decades of challenging weather conditions and armed conflict over the years and are now known for the high-quality organic strawberries that they grow. Like others in the Rincón Grande Cooperativa Rural Integral that they helped found, the Peréns are an indigenous Kakchiquel-Maya family who were granted their small farm during the agrarian reform of 1952, when huge tracts of land were redistributed from the hands of plantation owners to landless peasants. Their farm was part of a coffee plantation redistributed among the people of Rincón Grande.

Alongside their strawberry fields the Peréns also grow beans and squash in a traditional milpa for their own subsistence. Milpas are a millennia-old agroecological model, common throughout Guatemala and Mexico. They maintain soil fertility through a complex system of companion planting, allowing specific plants and herbs to protect or enhance the growth of others, either by attracting pollinators, repelling pests or acting sacrificially to detract pests from other plants.

Today the whole family works in the strawberry fields: Angela, the father Pedro, younger son Ulises, eldest daughter Lilian and sometimes the younger daughter, Nataly. The eldest son, Edmer, lives and works as an agricultural worker in British Columbia, Canada, and he helps support the family with the remittances he sends. While the Peréns struggle with ecological and economic burdens that bear down on many small-scale farmers in the area, Pedro tells us that they always remain positive, because lifetimes of resilience have shown them that there is always a way forward.

Photographer
James Rodriguez

LA MARINALEDA, ANDALUCIA, SPAIN

In this remote part of Andalucía, the town of Marinaleda has taken hold of its destiny. Without land or food and fed up with decades of poverty, in the 1970s the residents rose up and laid claim to the abandoned lands of the local nobility. They were led by their revolutionary mayor – Sanchez Gordillo, who has been dubbed a modern-day Robin Hood. After many years of protest, occupations and hunger strikes, in 1991 the town finally acquired 1200 hectares and started to plant the fields with food, such as peppers, artichokes, fava beans, green beans and broccoli. The farm includes 352 hectares of olive trees, which when harvested are turned into olive oil in the town's purpose-built factory.

Andalucía has one of the most unequal land distributions in Europe, with 2 per cent of families owning 50 per cent of the land. The large landless population had to survive on a few months of inconsistent seasonal farm work, handed out by the local landowners. The picture today is not too dissimilar. Since the housing bubble burst in 2008 with the economic crisis, abandoned beach resorts scatter the coast and the construction workers have been left jobless. Once again, Andalucía is seeing an exodus of its people to other parts of Spain and the rest of Europe.

In the midst of this, La Marinaleda offers an alternative social model that has proved its resilience and success. From 63 per cent unemployment in the 1970s, joblessness has become a thing of the past. Going against the neoliberal logic of 'efficiency', Marinaleda chooses to produce labour-intensive crops and reinvests all profits to create more jobs. Being the pioneers of organic agriculture in the region, they produce and sell everything, from organic olive oil to peppers and artichokes. For six and a half hours' work a day, the workers earn €47, which is twice the Spanish minimum wage.

Marinaleda is living proof of their motto, 'Otro Mundo es posible' – 'Another World is Possible'. The ecological, social and financial wellbeing of the community is at the heart of everything they do. Residents all enjoy free wifi, a communal swimming pool, gym and sports clubs, as well as free evening education. The locally built houses cost their residents €15 a month.

Inspired by other revolutionary leaders like Ché, Ghandi and Kropotkin, Mayor Sanchez Gordillo emphasises that 'Land should be a public good for the community who use it for producing food, for work and for the wellbeing of all.'

Photographer
Spencer Murphy

MARINALEDA
UNA UTOPIA HACIA LA PAZ

THE SEVERN PROJECT, BRISTOL, ENGLAND

Steve Glover doesn't come from a long line of farmers. He isn't following in the footsteps of his forefathers. Rather, Steve taught himself how to grow food following years of substance abuse. Having trained as a drugs counsellor, he knew first-hand that anyone recovering from addiction needed something to focus on, ideally something outdoors and motivating. He decided the best way he could help others coming out of treatment programmes was to connect them with the land and food, and provide a much-needed purpose within the community.

With a small grant, he took over an abandoned piece of land next to Bristol train station and with a team of apprentices, also recovering from substance abuse, he built his first polytunnel. Known as the Severn Project, Steve and his colleagues turn out a range of herbs and salads for restaurants and shops right across the city.

Steve points out why the Severn Project has been so important for so many recovering substance-abusers in Bristol: 'People leave rehab full of hope and plans for abstinence, but because they go back to their original environment, they start using again. They want to feel they have a purpose and a role in the community. Also, former drug addicts need exercise. If you want to stimulate your serotonin system in a non-sustainable way, take ecstasy. If you want to stimulate it in a sustainable way, then get some exercise!'

Since it started in 2010, the project has expanded, taking on another site in south Bristol with many more polytunnels, a refrigerated store and 22 delivery vans. The overheads are kept low by helpful donations such as compost from the Police, confiscated from cannabis factories around the city. Somewhat ironically, Steve has now diversified into growing industrial hemp, under a licence from the Home Office, which gets juiced and sold to healthfood stores as a treatment for stress and anxiety.

Steve is proud of the transformation he's seen in people: 'If you spend 15 years thinking about heroin and living in squats, when you finally come off and go into the big wide world you literally do not know how society works. You think *"what are all these people doing? Where are they going?"* After a year of working in an organisation like the Severn Project, they begin to understand what it's like to be a productive, useful member of the community. So instead of saying "I don't fit in", they walk down the street and say "We sell salad to that place".'

Note: Sadly, the Severn Project has since closed, but while in action it was a great inspiration and a safe haven for many.

Photographer
Martin Parr

THE GAIA FOUNDATION

For over 35 years The Gaia Foundation has been working to protect and revive the diversity of our living planet and the eco-centric cultures of Earth's best custodians - local and Indigenous communities.

Our vision is to restore a respectful relationship with the Earth and our approach is holistic and ambitious. At the grassroots, we work alongside local and indigenous communities to build back knowledge and resilience, strengthen self-governance and revive healthy ecosystems from forests to oceans. We build alliances and campaign to change and strengthen regional and global policies for protecting our planet and to respond to urgent threats. Through film, photography and other creative communications, we seek to reawaken peoples' sense of belonging and responsibility to a vibrant web of life.

Gaia's programmes in the UK and globally, from the Amazon to Africa to the Arctic, focus on: supporting small farmers, especially women, to enhance their traditional knowledge and seed varieties, to be food sovereign and to safeguard diversity for generations to come; protecting and restoring sacred landscapes and forests, and working with elders to secure their rights as traditional custodians of ecological wisdom and governance; backing communities and social movements at the front line of mining struggles to defend their ecological and cultural heritage and build regenerative futures; and spreading the practice and understanding of Earth Jurisprudence, through immersive trainings that support the revival and legal recognition of Earth-centred governance.

Find out more and support our work at **www.gaiafoundation.org**.

Opposite:

Zarraga, Panay Island, Philippines (*see* page 212)
Photographer Hannah Reyes Morales

THE PHOTOGRAPHERS

Kate Peters is a London-based photographer whose portraits form part of the permanent collection at the National Portrait Gallery in London. Kate visited the Walronds at harvest time in 2017. She says the experience reconnected her with her upbringing in the countryside and 'made me realise how important it is to experience nature. This, along with seeing people working so harmoniously with the land and raising livestock, made me question my lifestyle and what's important.'

Laura Hynd is a photographer and world explorer, best known for her projects 'Lady Into Hut' and 'The Letting Go'. About meeting Guillermo, Laura says 'I will never forget the moment Guillermo came through the gates of his farm to greet us. A wide, warm, beautiful smile accompanied by open arms. I instantly loved him! Scarf, hat, dungarees and his rescued 3-week-old pigeon sitting contented on his shoulder, Guillermo truly loves the land and all that grows there, be it fruit, vegetable, insect or animal. He affectionately kisses his chickens who trust him implicitly, talks to his horse, donkey and their mule with love and speaks of his land with a passion unmeasured'.

Influenced by the native Ugandan youth and hip-hop culture, **Kibuuka Mukisa Oscar** is a self-taught photographer and artist living and working in Uganda. He has been nominated for four awards and his work has been published in *The Washington Post* and *Art Base Africa*, among others.

Pieter Hugo is a photographic artist living in Cape Town. He has exhibited internationally for many years, participating in numerous group exhibitions at institutions including Tate Modern, the Folkwang Museum, Fundação Calouste Gulbenkian and the Bienal de São Paulo. His work is represented in the Museum of Modern Art, V&A Museum, San Francisco Museum of Modern Art, Metropolitan Museum of Modern Art, and the J Paul Getty Museum. Hugo received the Discovery Award at the Rencontres d'Arles Festival and the KLM Paul Huf Award in 2008 and was shortlisted for the Deutsche Börse Photography Prize 2012.

Stefan Ruiz was creative director at *COLORS* magazine. His work has appeared in *The New York Times Magazine, The New Yorker, Financial Times Magazine, Vogue* and *Rolling Stone*. Ruiz has exhibited widely including at the ICP New York, the Photographers' Gallery, London, Les Rencontres d'Arles, PhotoEspaña, Havana Biennial and the Contact Photography Festival in Toronto.

Jason Taylor is a photographer and filmmaker who is now based in the UK, after many years of living and working in India. He met Debal during this time and they became friends. He says 'When it comes to the science and conservation of indigenous foods and the protection of our fragile ecosystem, Debal is possibly one of the most important people working in these areas.'

Opposite:

Gerald Miles, Caerhys Farm, St Davids, Wales (*see* page 94)
Photographer Clare Richardson

Jo Ractliffe is an internationally acclaimed, multi-award-winning photographer whose most recent solo exhibitions include 'After War, The Aftermath of Conflict: Jo's Photographs of Angola and South Africa', Metropolitan Museum of Art, New York (2015); and 'Someone Else's Country', Peabody Essex Museum, Salem, Massachusetts (2014). Jo describes her experience at Shashe: 'In the week that I was there, I experienced an open-heartedness, a willingness to engage and debate, and a knowledge of farming practices that turned every preconception I might have had on its head. My conversations, particularly with the five women beer-makers, had an intimacy not unlike that which I have with friends at home. When I came home I started a small vegetable garden, installed two irrigation tanks and made plans to go back as the planting season began again.'

Martin Westlake is a British photographer who lives in Indonesia. His work has been published in *Monocle*, *Wallpaper*, *The Telegraph*, *The Sunday Times* and other publications. Martin visited East Flores at harvest. He says: 'Maria's passion for sorghum and unselfish support of the farmers made an indelible impression on me. For years, she has gone against the Indonesian government which culturally shames those who eat sorghum. I hope that as the success of the sorghum farmers of Likotuden becomes more visible to a global audience, the Indonesian government's policies and attitudes towards sorghum-farming in remote island communities will change.'

Sophie Gerrard is an award-winning photographer who specializes in contemporary documentary stories with environmental and social themes. She has achieved over 45 exhibitions and publications, been nominated for 18 awards and is a lecturer in photography at Edinburgh Napier University.

Spencer Murphy is a Fine Art and Commercial Photographer who lives and works in London, contributing to many magazines, including *The Guardian Weekend*, *The Telegraph Magazine*, *Time*, *Monocle* and *Wallpaper*. His portraits have also appeared in publications such as the *Rolling Stone*, *GQ* and *Dazed and Confused*. In 2013, he won the National Portrait Gallery's Taylor Wessing Photographic Portrait Prize.

Fabrice Monteiro is a photographer whose images question 'the evolution of black identity through history'. Engaging with issues of sustainability and mythology, his aesthetic taps into photo-reportage, fashion photography and traditional studio portraiture. Monteiro's work has been exhibited in Belgium, Gabon, Spain, Germany, France, the US and Luxembourg. His photographs are in the permanent collections of the Chicago Museum of Photography, the Seattle Museum of Art and the Geneva Museum of Ethnography, Switzerland.

Omar Victor Diop is a fine art photographer and is in demand by galleries and advertisers alike. He was born in Dakar, and his work is described as interrogative and intriguing, prospective and uplifting, while drawing on his African visual heritage. Omar travelled to Kenya to meet the women on their farms. He then created these artworks, which combine vintage botanical illustrations from the early 1900s with photography. The fruits, cereals and vegetables that are inserted in these images are part of what Agata, Lucy, Margaret, Dionisio, Magrine, Basilia and Mercy grow in their farms all year long, thanks to the reintroduction of crop-rotation techniques.

Susan Meiselas has been a Magnum photographer since 1976, when she became known for her documentation of the Sandinista Revolution in Nicaragua. She directed two feature films about the revolution. She has had solo exhibitions in London, Paris, New York and Los Angeles and received the Robert Capa Gold Medal for her work.

Clare Richardson is a photographer and manages a family farm in Wales. She has exhibited at the White Cube Gallery and the V&A in London and published books with SteidlMACK. Her most recent book, with Antony Gormley and Jeanette Winterson, is called LAND. She had heard about Gerald before actually meeting him, having enjoyed many a wholesome meal made from his produce. She says: 'Larger than life, and full of energy, you can see how Gerald has managed to galvanize and help mentor many others into setting up CSAs in the UK. If we could scale up the concept of Community Supported Agriculture, I think we would have a food-supply chain that would benefit all.'

Michel Pou is a Cuban photographer from Havana. He has worked for Estudio 50, Titina, covered events such as 'Havana World Music' and featured in exhibitions in the US, such as 'El Yuma: Looking at the US from the Island', and 'Ola Cuba' in France. He is interested in documenting his Cuba, the changing times and has photographed many Cuban farmers who are part of the Cuban agroecological renaissance.

Sian Davey is an awarded photographer in the UK, who investigates the psychological landscapes of both herself and those around her. She has been involved in 13 exhibitions and nominated for 22 awards internationally. Sian says 'I was amazed and in awe at what both Dee and Adam were achieving. I am not sure how they do it, but they do and perhaps that's the point. Their love and commitment to farming and growing food locally is very possible, whether it's in acres or pots in your backyard. When you grow your own food your relationship with the world around you strengthens and you generally feel so much better.'

Born in Afghanistan, **Zalmaï** left the country after the Soviet invasion in 1980. He travelled to Switzerland, where he became a Swiss citizen. Zalmaï works as a freelance photographer and has been published in the *The New York Times Magazine*, *Time Magazine*, *The New Yorker* and *Harper's Magazine*, among others. He has also worked for a number of international organisations and NGOs, including Human Rights Watch, International Committee of the Red Cross, UN Office On Drugs and Crime, and the UN Refugee Agency.

Carolyn Drake is an American photographer and an associate at Magnum Photo. She is the recipient of a Guggenheim fellowship, the Lange-Paul Taylor Prize and a Fulbright fellowship, among other awards. She says she is drawn to photograph people who resist domination and was interested in the Masumotos for this reason.

Davide Degano graduated from the Royal Academy of Art (KABK), in Den Haag, and works as an independent photographer. He grew up in northern Italy and is related to the Perabò. family. He says 'I never paid much attention to how my environment shaped me until I went to photograph the Perabò family. In doing so, I rediscovered the importance of feeling connected to nature, to be a part of it, not just standing above it. It has taught me to respect other forms of life, and to be patient and consistent in what I do. I now appreciate being a "dude from the country!"'.

Jordi Ruiz Cirera is an independent documentary photographer who has won numerous international prizes including the Taylor Wessing Photographic Portrait Prize and the Magnum Emergency Fund Grant. He has exhibited widely in galleries and festivals, and is a regular contributor to media such as *The New York Times*, *The Sunday Times Magazine* and *The Guardian*. Jordi spent the day with Remo and Irmina and their family as they celebrated one of the grandchildren's birthdays. Jordi says 'Remo explained to me how the current agroindustry works against human needs, in creating necessities where there are not, and using a huge amount of energy and land that could be used otherwise. What he and the family wanted to create was a place of change, where society could be re-imagined in a more sustainable way.'

Tina Hillier is an award winning portrait and documentary photographer based in the UK. Her work has been widely published and exhibited in group shows most notably at the Bargehouse Gallery, Portcullis House and at The National Portrait Gallery. Tina says 'It was an enlightening experience spending time with a woman of my age leading a completely different life. I was envious of her daily connection with nature and her knowledge of agriculture; life skills once fundamental to our existence, now redundant in most city lives.'

Zhang Kechun is a Chinese artist and photographer who lives and works in Chengdu. He is known for his photographs dwelling on the significance of the landscape in modern Chinese national identity. He won the National Geographic Picks Global Prize in 2008, was nominated by Sony World Photography Awards in 2012 and 2013 and was nominated for the Magnum Photo Awards in 2017.

Antoine Bruy is a French photographer and winner of many awards including this year's Emerging Photographer Fund by *Burn Magazine*. Antoine's projects have included Scrublands, which focused on people living self-sufficiently. His photographs have been featured in numerous publications worldwide including *The New Yorker*, *The Washington Post*, *The Guardian*, *WIRED*, *Slate*, *The Huffington Post* and *Le Monde*.

Niall O'Brien is an Irish photographer who has exhibited widely on a number of continents. He was shortlisted for the Fuji Photographer of the Year award in 2003 and was the Selected Photographer for the Singapore International Photography Association. Niall describes his trip to Santa Teresa: 'I was met by an incredible team of farmers who were very proud of the work they do and were eager to help me understand how positive the co-op is to them and the farms they care so much about. All the farms seemed to thrive and the quality of the coffee was extremely important to them.'

Nick Ballon is a documentary and portrait photographer based in the UK. For the last decade Nick's personal work has focused exclusively on his Anglo-Bolivian heritage, exploring socio-historical ideas of identity and place. He has won over 12 awards including the Best in Book and Personal prize at the 2017 CR Photography Annual.

Rankin is a British photographer, publisher and film director. With a portfolio ranging from portraiture to documentary, he has shot The Rolling Stones, David Bowie, Kate Moss, Kendall Jenner and The Queen to name only a few. Fearless behind the lens, his imagery has become ingrained in contemporary iconography. As both a photographer and director, he has created landmark editorial and advertising campaigns for some of the biggest and most celebrated publications, brands and charities. Rankin lives in London with his wife Tuuli and their dogs.

Russell Kleyn is a photographer based in Wellington, New Zealand. His portraits are included in the permanent collections of the National Portrait Gallery in London and the New Zealand Portrait Gallery in Wellington. Russell says 'The Māori meaning of Arohanui is "with deep affection" or "big love" and it was with this open-hearted spirit that Aunty Hanui welcomed me onto the Waipatu Marae and treated me as part of their community when photographing at Aunty's Garden.'

Rena Effendi is a social documentary photographer, whose early work focused on people's lives in Azerbaijan, Georgia and Turkey. Her work has been exhibited worldwide and she has worked on assignments with *National Geographic*, *The New Yorker*, *Marie Claire* and more.

Graciela Iturbide was born in Mexico City, where she continues to live and work. Iturbide has exhibited widely, including solo exhibitions at the Centre Pompidou, MOMA, San Francisco and Barbican Centre. In 2013, she had a Retrospective at Tate Modern. Iturbide has received numerous awards, including the Hasselblad Award in Sweden, and her work is held in many public collections including MOMA, New York; MOMA, San Francisco; and the Museum of Modern Art, Mexico City.

Joel Karppanen is an award-winning photographer and artist from northern Finland. Well known for his socially and historically aware works, he concentrates on making personal, long-term documentary photography series and essay films. Joel has received recognition including a New Photo Journalist Award and Young Hero Grant in 2017. His first monograph *Finnish Pastoral* was published in 2018, his first museum solo show was held in the Aine Art Museum in 2019, and his works have been exhibited in London, Vienna, Bratislava and Helsinki among other cities. Fascinated by the views and the atmosphere of Selkie, he says 'I was inspired by the sincere connection the community had with nature.'

Pat Kane is a journalist and photographer in the Northwest Territories covering the far north of Canada. He is a regular contributor to many magazines including *Monocle*, *VICE*, *The Week* and *Canadian Geographic*. About his experience at Tathlina Lake he says, 'I was a stranger to the residents of Kakisa but was welcomed with open arms by my hosts at Tathlina Lake where we camped for nine days. It was an incredible experience. We shared chores, hunted, ate food together and told each other about our lives. The people of Ka'a'gee Tu First Nation are so open to sharing their stories and rich culture. I feel lucky to now call them my friends.'

Tanya Habjouqa is a documentary photographer with a primary interest in gender, social, and human rights issues in the Middle East. She is the author of *Occupied Pleasures,* heralded by *TIME* magazine and the *Smithsonian* as one of the best photo books of 2015 (winning her a World Press Photo award in 2014). Tanya is a founding member of *Rawiya,* the first all-female photo collective of the Middle East and she is currently based in East Jerusalem.

Jane Hilton is a London-based photographer who takes portraits in the Midwest of America. She says her work is about the extraordinary realities of ordinary people's everyday lives. She turned one of her portrait series about a brothel in Nevada into a ten-part BBC series. Her work has appeared in numerous major publications such as *The Sunday Times Magazine*, *The Telegraph Magazine* and *Financial Times Magazine*.

Andrew Esiebo's works have been exhibited worldwide and published in media publications including *National Geographic*, *The Guardian*, *The New York Times* and *Time*. He is a 2011 Musée du Quai Branly artistic-creation prize winner. About his visit to Burkina Faso, he says 'In my many years of working in Africa, I have never seen women in a community who are so empowered and happy to speak out. The women I photographed had a voice in their community, in fact stronger voices than many of their husbands! They have gained a lot of confidence and economic independence from working together to grow food and as a result they have been able to buy their families better houses and put their children through school. When I arrived they were really excited to show me how much their produce has improved – from four bags of cereal to over 15.'

Hannah Reyes Morales is a Filipina photojournalist whose work focuses on individuals' complex situations created by inequality, poverty and impunity. Her work has featured in *The New York Times*, *The Washington Post*, *Wall Street Journal*, *Time*, *National Geographic*, *The Guardian*, and *Lonely Planet*, and has been exhibited in Manila, Telluride, Copenhagen, Aalborg, Nanning, Suwon and Chiang Mai.

Katrin Koenning is a photographer from the Ruhrgebiet, Germany, who regularly exhibits her work in solo and group shows in Australia and internationally. Her work has been published widely in *The New York Times*, *Financial Times Magazine*, *National Geographic*, *The Guardian* and others. Colin told her 'It was the fury of the fire that taught him to listen to the land, and to instigate its slow path back to healing through sustainable methods. We talked about so much it was hard to say goodbye to him at the end of that day.'

Matt Eich is a photographic essayist working on long-form projects related to memory, family, community and the American condition. He is an Assistant Professor at Corcoran School of the Arts and Design at The George Washington University in Washington DC and the author of four monographs of photography. Matt publishes under the imprint Little Oak Press and resides in Charlottesville, Virginia.

Clare Benson is a photographer, interdisciplinary artist and educator based in New York City. She has worked alongside hunters, herders and space scientists, exploring connections across systems of nature, tradition, science and mythology. Her work has been featured in exhibitions and screenings across the US and internationally.

Federico Pardo is a Colombian biologist, photographer and documentary filmmaker. He is a National Geographic Explorer and the winner of two Emmy awards: one for best documentary cinematography with National Geographic's *Untamed Americas* and another one for the outstanding feature story in Spanish, *La Amazonía: Un Paraíso a La Venta*. Federico's wildlife and documentary photography has been published widely both in print and digital media and his personal work has been exhibited in galleries in Bogotá, México City and Madrid.

Holly Lynton is an American photographer whose work explores people's passion for maintaining rural traditions and preserving natural resources despite the challenges of agribusiness, climate change and technology. Her series 'Bare Handed' has been exhibited in several solo exhibitions in cities in the United States, including Miami, Denver, Boston and Chicago, as well as internationally. Holly went back to visit Shannon several times, on the last occasion taking home two fresh squid for her family. '"So why do you do it?" I asked her as we'd talked about the decrease in fish. "I like feeding people," she said. The ability to haul food from the sea, without taking in any waste, any bi-catch, and then being able to supply her community with sustenance was all the motivation she needed to persevere.'

Lindeka Qampi is a self-taught photographer who has focused her lens on daily township life, with particular attention on Khayelitsha, the township in which she has lived since her teens. Since 2012, Qampi has been the project facilitator for Inkanyiso, an activist platform founded by fellow photographer, Zanele Muholi. Together they have held visual workshops in Benin and Norway and in 2016, were acknowledged for their work with a Brave Award. Qampi has exhibited in Cape Town, Italy, Norway and the US.

Bruno Morais is a Brazilian photographer and founder of the Colectivo Pandilla in Rio de Janeiro. Colectivo Pandilla has shown his work at the gallery Vitrines during FotoRio, Gallery 535 as part of the Favelas Observatory, at FB Gallery in New York City and at the National Historical Museum in Rio de Janeiro. After spending three days with communities of the Movimiento de los Trabajadores Rurales Sin Tierra (MST), Brazil's Landless Workers' Movement, he said: 'The power of MST comes straight from its grassroots base, the political consciousness of its many members. The movement gives them something that surpasses any material difficulty – dignity.'

David Chancellor is a multi-award-winning documentary photographer. Named Nikon Photographer of the Year three times, Chancellor also won the Taylor Wessing National Portrait Prize in 2010, received a Sony World Photography Award (Nature and Wildlife) and the Veolia Environment Wildlife Photographer of the Year Award.

Jack Latham is a Welsh photographer based in the United Kingdom. His work is often presented in a mixture of large-format photography and filmmaking and has been realised in several self-published books. Latham is the author of *Sugar Paper Theories*, 2016, which was shortlisted for the Paris Photo/Aperture and Krazna Krausz photobook awards. The exhibition of the same name was later longlisted for the Deutsche-Börse Photography Prize 2017. His work has been exhibited and published internationally.

James Rodríguez is a documentary photographer who was raised in Mexico City. His work primarily focuses on Guatemala and has been published in *The New York Times*, *Der Spiegel*, *Le Monde*, *The Guardian*, among others. He has also exhibited in numerous venues worldwide, including the Corcoran Museum in Washington DC and Mexico City's Museum of Memory and Tolerance.

Martin Parr is a chronicler of our age, who takes an intimate, satirical and anthropological look at aspects of modern life, in particular documenting the social classes of England. He has published 40 photobooks and had over 80 exhibitions around the world including his internationally touring 'Parrworld' (2009). He has curated many exhibitions, among them Arles Festival, Brighton Photo Biennial and New Typologies, New York Photo Festival. He joined Magnum Photos in 1994 and was the President of Magnum Photos International from 2014 to 2017.

CACAO

ACKNOWLEDGEMENTS

We Feed the World was the ambitious shared vision of Francesca Price, Project Director, and Cheryl Newman, Curator. This book honours their determination, commitment and hard work in bringing it to fruition alongside The Gaia Foundation team.

Together we are indebted to the talent and generosity of some of the world's most respected photographers and the spirited response from farmers, fisherfolk and communities that we approached to feature in this project.

We would like to acknowledge and thank the many global partners who were involved in the stories and photo shoots, and who work tirelessly every day in support of regenerative agriculture and local food systems that nurture the Earth. From all continents, these partners include Bhutan Network, Canaan Palestine, Colectivo por la Autonomía, Community Supported Agriculture, Network, Elkana, Food Ethics Council, Fundación Entre Mujeres, Gaia Amazonas, Global Greengrants, Groupe de Recherche et d'Action pour le Bien-Etre au Bénin, Greenpeace China, Groundswell International, GRAIN, Hopi Cultural Preservation Office, Kawsay, International Tree Foundation, La Marinaleda, Landworkers Alliance, La Via Campesina, London Mining Network, Movimiento de los Trabajadores Rurales Sin Tierra, Muonde Trust, Northwest Atlantic Marine Alliance, Red Andaluza Semillas, Red en Defensa del Maiz, Riverford, Samdhana Institute, Scottish Crofting Federation, Slow Food Uganda, Slow Food USA, Soil Association, Snowchange Cooperative, Sustain, Tides Canada, Vivero Alamar Coop, Yes to Life No to Mining, Zarraga Integrated Diversified Organic Farmers' Association, Zaytoun and the Zimbabwe Smallholder Organic Farmers' Forum.

The A Team Foundation believed in this project from the start, and we are eternally grateful for their wisdom, support and funding throughout. We acknowledge them for the success of the project and publication of this book, along with the Agroecology Fund, New Venture Fund, Casey and Family Foundation, Martin Stanley, Natracare and those who gifted anonymous but critical donations.

Other generous funders and small ethical businesses contributed to making the *We Feed the World* project such a huge success from the flagship exhibition at the Bargehouse on London's Southbank, to small, local exhibitions around the world.

Opposite:
The Cecaq-11 Cooperative,
São Tomé Island, Africa (*see* page 258)
Photographer David Chancellor

With huge gratitude to the Bertha Foundation, the Big Lottery Fund, Café Direct, Craignish Trust, Crepes and Waffles, Divine Chocolate, Dr Bronner's, Giant Peach, Greenhouse PR, Halleria Trust, My Green Pod, Neals Yard Remedies, Nutiva, Panta Rhea Foundation, Petersham Nurseries, Rebel Kitchen, Roddick Foundation, R. Steiner Foundation, Savitri Waney Charitable Trust, SAHARA, Sheepdrove Trust, Tide Canada Foundation and Yeo Valley. We could not have done it without you!

We are grateful to the many *We Feed the World* ambassadors who entrusted their wisdom and actively helped to promote the project. Special mentions to Anna Lappé, Anna Van Der Hurd, Arizona Muse, Ben Raskin, Dame Vivienne Westwood, Emily Kerr-Muir, Hugh Fearnley-Whittingstall, Jeremy Irons, Jessica Sweidan, Lara Boglione, Robert Reed, Sam Roddick, Tom Hunt, Vandana Shiva and Vic Coppersmith-Heaven.

Thanks to pro-bono legal advisors John Enser and Louise Keenan. Huge praise for the exhibition design work by Dom Carroll. And appreciations to Little Toller, the wonderful publishers bringing this book to the public and themselves reflecting the small-business ethos that resonates with so many of the farmers featured.

And finally, we salute The Gaia Foundation team, our Trustees, and the many enthusiastic interns and volunteers, who, over a period of more than three years, have given their heart and souls to bringing these stories and images to the world. With special mention and thanks to Dijana Malidza, Fiona Wilton, Georgie Styles, Hannibal Rhoades, Iara Monaco, Liz Hosken, Rachel Karniely, Rowan Phillimore and Thomas Takezoe.

Francesca Price (Project Director) is a journalist, campaigner and the founder of the *We Feed the World* project. She spent three years interviewing the communities who feature in this book and launched the exhibition in London in 2018. For the last ten years she has worked with the global food and farming movement, finding creative ways it can communicate with larger audiences. She is currently the Programme Director of the Oxford Real Farming Conference Global.

Cheryl Newman (Artistic Curator) is an artist and independent curator of photography living in London. She recently completed and MA in Photography Arts at the University of Westminster (Distinction). Her personal practice explores her history though archive and family images and the revisiting of memory and belonging. In addition to *We Feed the World*, her recent curatorial projects include; 'Time to See', The Queen Elizabeth

Diamond Jubilee Trust: Exhibition and book, focusing on avoidable blindness within Commonwealth countries and '209 Women', one of the highest profile exhibitions of 2018 in which all the female MP's in the UK Parliament were photographed by women photographers to celebrate 100 years of suffrage. This moved to Open Eye Gallery Liverpool in February 2019, and is now part of the Parliamentary Art collection. For more than fifteen years she was the Photography Director of the award-winning Telegraph Magazine where she raised the profile of the magazine and commissioned intelligent and inventive photography worldwide. She is a nominator for the Deutsche Borse Photography Prize Foundation and serves on juries globally, including; The Moran prize for Contemporary Photography, Sydney, Australia, 2015—2019 and 'The Eugene Smith Grant', jury 2019. She was the chair of the RPS Photography Awards 2017—2019. Cheryl is also an educator, runs the Dartington Workshops with artist Sian Davey and an ongoing mentor programme in Oslo, Norway.